Brew from scratch

Brew from scratch

James Morton

slow down, make beer

photography by andy sewell

Hardie Grant

QUADRILLE

Contents

Love beer

What is beer?

Beer: broken down

Water, malt, hops, yeast;
Mash, boil, cool, ferment

Beer is made from **water**, malted barley ('**malt**'), **hops** and **yeast**. You could add loads of other things, such as wheat, rye, oats, rice, corn, sugar, honey, spices, fruit, fruit juice, and you've still got beer.

The first stage of brewing is when hot **water** and **malt** are steeped together in a 'mash tun' – this process is called '**mashing**'. Mashing is done at precise temperatures, in order to activate enzymes that break down the starches in the grains into sugars. After an hour or so, most of these sugars will have dissolved in the hot water to give a hot, sticky liquor called 'wort' (pronounced 'wurt').

The next step is **lautering,** which is to remove the spent grain and leave only the wort. The wort is drained out of the mash tun and into a separate large tank called the boiler, or kettle. The remaining grains are then rinsed with hot water to remove any remaining sugars – **sparging**, my favourite word – and this weaker wort is added to the kettle.

This is brought up to a **boil**, and it's boiled for an hour or so, during which time **hops** are added at various intervals. Hops are added for flavour and aroma only. They give bitterness if they're added near the start of the boil and aroma if added near the end. Hops don't provide any more sugar and so they do not make the final beer stronger. Traditionally, hops have been used as a preservative, as they slow down the growth of some bacteria that might infect the beer.

After the boil, beer is **cooled** down to roughly room temperature and transferred to another large tank – the fermenter. Here, **yeast** is added. Which yeast you choose defines whether you've got an ale or a lager; this book is concerned largely with ales. Over the next few days to weeks, the beer undergoes **fermentation,** where the yeast turns the sugars into alcohol, among other delicious things. The more sugar there is to begin with, the more alcohol will be produced and thus a stronger beer will be made.

After fermentation, sometimes more hops are added and sometimes the beer is aged or conditioned for long periods of time. Nowadays, most beers are pumped full of carbon dioxide to make them fizzy and then transferred to kegs or bottles to be enjoyed by us, the paying public.

Some beers are **bottle conditioned**. This means the beer is put straight into bottles, without any forced carbonation. Instead, sugar is added, which can be eaten up by the remaining yeast, or by additional yeast if the beer was filtered or fined. This causes the pressure to build up inside and results in a fizzy final product. Add too much sugar, though, and you could have bottles with the potential to explode. And when they go, they do go. **Cask conditioning** works in exactly the same way; just think of a cask as one big bottle. This is why hand-pulled 'real ales' tend to have such a short life – once the first pint has been pulled, it's just like having opened a bottle and poured a little from the top. Every pint from then on will be different from the last.

When we make beer at home, we can go through exactly the same processes, using smaller equipment. Size is the only difference between home and commercial brewing. This allows us to make any beer at home to the same quality as you can buy down the pub.

A note on measurements

My dear American cousins: one day you'll come to love the metric system, at least in part, as we now do in Britain. To facilitate this transition, I recommend you tackle every recipe in this book with a set of good digital scales and in grams. Your accuracy and thus your brewing ability will improve. Other quirky 'Britishisms' can be deciphered through a quick Google. As a helpful starter, 'cling film' is what you call saran wrap.

What beers should taste like

As soon as you start to talk about beer 'styles', the brewing world becomes divided. The question that's usually asked is about whether beers *should* fit into certain styles. On one hand, amazing beers have been made within certain parameters for generations, and thus it stands that operating within these same standards will result in good beer. On the other hand, many argue this limits creativity and diversity, and people should be able to make beer any way they want.

How to taste beer
Before you taste your beer, **pour** it into a glass. **Look** at it – is it dark or light, clear or cloudy? How's the head? Then **smell** it. Get your nose in there and give it a good sniff. You'll get some burn from the CO_2, but that's OK – sniff a good two or three more times. Then you can **taste**. But don't sip; swig. Turn the beer around your mouth for as long as is socially acceptable. Don't forgot to scowl, frown and then nod with faint approval.

Some styles to seek out

Here are a few styles to find and to try, and to taste. It's not an exhaustive list. I can, however, impartially recommend such a text: for an excellent, detailed and comprehensive list of beer styles and what they involve, check out the Beer Judge Certification Program (BJCP) Style Guidelines at bjcp.org. They're free, and put together not for profit. This allows you to see what they're looking for when you enter your beers into competitions.

British and Irish ales

English pale ales, bitters and IPAs
These begin with 'golden ales': pale, thirst-quenching beers best enjoyed at an English summer music festival. They tend to be very dry, weak, very drinkable and not hugely hoppy, with what flavour there is coming from English hops and yeast. IPAs (India Pale Ales) are stronger, hoppier pale ales. Bitters are similar, but have a reddish hue and lingering sweetness from caramel malt.

Stout and porter
These beers are noticeable for their dark colour and coffee or chocolate-like flavours, caused by the addition of roasted malts. They can range from very sweet (a 'milk stout') to silky smooth (an 'oatmeal stout') to very dry and drinkable (an 'Irish dry stout': think Guinness). As they go up in strength, they tend to become even darker and sweeter. The strongest are Russian imperial stouts, which can be anything from 7% to 20% abv plus.

Old ale, barleywine and Scotch ales
On old ale tends to be aged, roasted and slightly sweet, with a pleasant 'stale' quality. The Scotch ale ('wee heavy') is a hugely sweet, strong (7–10% abv) and nutty beer. The English barleywines do tend to be stronger and richer, with intense and complex flavours that reveal themselves with a good bit of aging. Expect them to be 8–12%.

Belgian beers

Belgian pale and blond ales
These styles are exceedingly simple beers that showcase the wonderful abilities of Belgian yeast strains. They tend to be nothing more than pale malted barley, minimal English or European hops and some sugar.

Saison
Saisons tend to be strong (but don't need to be), pale, refreshing and highly carbonated, with distinctive flavour resulting from distinctive saison yeasts. Watch out for a peppery spiciness and tangy aromas of freshly cut citrus fruit. Brewers (myself included) often add 'wild' yeast strains, such as *Brettanomyces*, to add some funkiness and get the beer as dry and refreshing as possible.

Abbey ales
Often produced by monks for humanitarian reasons, these beers tend to be strong with a deep caramel hue. They are exquisite and complex, with notes of raisin and caramel, as well as a whole bunch of spicy phenols and subtle fruity esters.

Lambic
Lambics: the champagne of Belgium. Sour, exceedingly complex and often with a healthy note of barnyard funk – these are some of the most divisive beers among new craft beer lovers, but the most rewarding for those who love them.

Germanic beers

Pale lagers
The lagers in the German style are the best lagers, and some of the best beers, in the world. They are only distant relatives of the production line 'macro' examples drunk daily by so many of us, and deserve reverence for the time and effort that's involved in making these beers. Ranging in colour and strength, pale lagers are known for their maltiness and hoppiness. Stronger ones are called 'Bocks'. For a pale, weak, exceedingly drinkable style, try a Munich Helles.

Dark lagers
Lagers don't always have to be straw-coloured – the Munich Dunkel can be as dark as any stout or porter, and filled with roasty, chocolate-type flavours. A lager should always be drinkable, though, and this beer is no exception. A Schwarzbier is a very similar style, though it hails from Saxony and tends to be even darker, drier and more roasty.

Wheat beer
Weissbier (German wheat beer; Hefeweizen) is instantly recognisable in both sight and smell. You'll see its huge foamy head and cloudy amber body, but then you'll smell two of the most distinctive yeast characteristics in all of beer: banana and clove. Dunkelweizens employ darker malts to give a more bready, caramelly or even roasted characteristic. Weizenbocks are much stronger versions, and can be dark or light.

American beer

American IPA
This is a bastardisation of the original English IPA, and what a beautiful bastard it is. This style, more than any other, has fuelled this new beer renaissance across the world. They took a traditional English IPA, refined it and added more hops. Awesome American hops. And lots of them, both at

the end of the boil and then after fermentation has completed, in a process known as 'dry hopping'. Solid examples, such as Lagunitas IPA or Stone IPA, will smell floral and citrusy and piny and blow you away with their hop flavour, before smacking you in the face with a hefty blow of bitterness. Their dry finish will make you go back for more and more.

New England IPA

This is the beer style of the 21st century. Think a standard American IPA, but with very few bittering hops and a massive dry hop. To use anything but Pale Malt, Oats and Wheat is sacrilegious, and finings are avoided. This means they're opaque, and dangerously drinkable.

American pale ale

This is a weaker, less hoppy and fuller-bodied IPA. It tends to be a bit closer to its English roots with a more malt-focused backbone, and not all of them will be dry hopped. They will still use plenty of American hops, though. Check out the classic: Sierra Nevada Pale Ale. An American amber ale is a darker and fuller version of the pale ale that uses more crystal or caramel malts – think of an English bitter, but with American hops.

Double IPA

A double IPA (imperial IPA) is a style pioneered by Vinnie Cilurzo of Russian River Brewing. His beer, Pliny the Elder, is still considered one of the world's best beers. It's a simple concept – an American IPA, but *more*. Stronger, hoppier, drier, bitterer. The key to their success is an exceedingly dry finish and unobtrusive malt flavour, letting the aromatic American hops shine through.

California common

This is a 'hybrid' ale that offers home brewers the chance to create a lager at normal, ale temperatures. Also known colloquially as a 'steam beer', after Anchor Brewery's original and defining example, this style predates refrigeration. The yeast was bred to produce lager characteristics in the cool-but-not-cold air of San Francisco. California commons are amber in colour, assertively bitter, with a fuller body than most lagers.

Beer colour chart

1	4	7	10	13
2	5	8	11	14
3	6	9	12	15

1   MUNICH HELLES
2   BELGIAN STRONG GOLDEN
3   SAISON DUPONT
4   AMERICAN IPA
5   ENGLISH BITTER
6   HEFEWEIZEN
7   OKTOBERFEST MARZAN
8   ORVAL
9   EPA
10   KRIEK
11   ENGLISH BROWN ALE
12   BELGIAN STRONG DARK
13   AMERICAN BARLEY WINE
14   IMPERIAL STOUT
15   OATMEAL STOUT

Equipment

1 Plastic buckets
2 Spray bottle
3 Airlock
4 Hydrometer
5 Capper
6 Bottle caps
7 Taps
8 Silicone tubing

This guide does assume you do have a couple of things – a set of **kitchen scales** and a **large spoon**.

Basic stuff

No-rinse sanitiser

I recommend Star San by Five Star Chemicals, or a similar copycat product. It will be your single best brewing investment. It might be a bit pricey, but it lasts for ages. Star San is a dual combination sanitiser that contains a surfactant and an acid. It cannot penetrate through grub and muck, so must only be used on clean surfaces.

To use, dilute 1.5ml/⅓ tsp of no-rinse sanitiser (I use a syringe for accuracy) in 1 litre/quart of tap water. I keep it in a **spray bottle (1 litre/ quart capacity)**, meaning I can sanitise most of my equipment with just a few swift sprays.

Thermometers

First, a **liquid-crystal thermometer strip** that sticks onto the side of your bucket is a necessity, unless you want to be opening it up all the time. Then, you'll want a **digital probe thermometer**. They're fast and accurate and easy to read, as well as being easy to clean and then sanitise with a few sprays of no-rinse sanitiser. They start very cheaply, but if you can, get one that's waterproof (we've all dropped these into a batch of beer).

Plastic buckets (25–30 litre/quart capacity) and taps × 2

A bucket is what your beer will ferment in – your primary fermenter. You need a lid to go with it and a handle helps you to carry it around. It needs to be food-safe and designed for brewing. Further desirable properties are translucency (for visibility when your beer is fermenting) and volume graduations, so you know how much beer you've got. Get two: an extra one doesn't take up any more space, as it should just slide inside the first.

A second bucket (with a lid) can be used as a *secondary fermenter*, in which you can add 'dry hops', and as a *bottling bucket*. This is a separate vessel into which you place a sugar solution for carbonating your bottled beer, onto which you decant your beer from the 'primary' fermenter.

Taps are just for ease of transferring your beer – they are a host for infection so take special care when cleaning them. You can buy buckets with holes pre-drilled for taps, or your home brew equipment supplier can offer to do these for you.

Airlock

An airlock is a device that acts to lock air out of your bucket. As the beer ferments, it produces carbon dioxide (CO_2). This causes an increase in pressure in your fermenter, and forces the air out through any available opening. This means that the *headspace* – the volume of the fermenter that's not filled with beer – becomes filled with mostly CO_2.

The airlock stops any air from getting back in and ensures that only CO_2 is in contact with your beer. I fill my airlocks with sanitiser, which acts as a barrier. Air would only be able to re-enter if the pressure inside dropped, which it isn't going to do, as your fermenting beer should constantly be releasing CO_2.

Hydrometer and plastic hydrometer jar

A hydrometer is a very simple device that measures the density of a liquid. It uses a scale called *specific gravity* (*gravity*, for short). A hydrometer looks like a round, pointy glass rod with lines on it, because that's all it is (see page 24). Depending on how much it floats or sinks in various liquids, you can tell how dense they are. It usually comes with a thin plastic case that can be used as a jar, but I recommend you buy a specific plastic hydrometer jar. Don't buy a glass one, as they break really easily. Get a plastic hydrometer jar if your chosen supplier has them cheap, but otherwise go online and buy a 100ml/3½fl oz measuring cylinder.

On most hydrometers, pure water at 20°C (68°F) will have a gravity of 1.000 (one point zero, zero, zero). Add sugar to the water, and its density will increase. Therefore, the *gravity* will increase. Adding alcohol to the same water will cause the gravity to go down, because alcohol is less dense than water.

As things heat up, they get less dense. So the same sugary water at 40°C (104°F) will appear to have a lower gravity on the same hydrometer. If the temperature of your beer is wildly different from 20°C (68°F), I'd recommend using an online calculator or app to calculate the true reading.

This all means we can track how well a beer is fermenting by taking an *original gravity* (OG) before we add the yeast – this tells us roughly how much sugar we've got to work with. Then, as the yeast turns the sugar into alcohol, the gravity gradually reduces until it reaches the *final gravity* (FG). This is as low as the gravity will get. Despite the alcohol content, the gravity will usually not go below 1.000 in beer. This is due to residual proteins and unfermentable sugars that yeast can't metabolise.

By knowing the OG and FG, we can find out how much alcohol has been produced and thus the alcohol content of our beer.

Silicone tubing (2 metres/80 inches)

You need tubing to move your beer from one place to another. Investing in good tubing is one of my top recommendations when starting out. Good tubing is thick enough that you can hold it even with boiling water running through. For your own comparison, my stuff has an inside diameter of 12mm/½ inch and outside diameter of 21mm/1 inch.

Bottling stick

This is one of the most underrated pieces of brewing equipment. It's basically a piece of hard plastic tubing with a valve on the end. Connect it to your bucket's tap, open it up and beer will flow into the bottling stick, but not out of the end. Then, when you place a sanitised bottle fully over the end of the tube, the valve hits the bottom of the bottle and fills the beer from the bottom up.

Try to get a stick that can attach straight on to your tap. If you can't find one, you can cut off a very short piece (5cm/2 inches) of silicone tubing. Use a *worm-drive clip* (metal hose clamp) to close the tube over the end of the stick. You can then push the tubing on to your tap.

Bottle capper, caps and bottles

There are two types of capper – bench cappers and twin-lever cappers. The former, despite being sturdy and fast, are large, expensive and require adjustment when using different-sized bottles. A cheap twin-lever, works fine. Get a few hundred 26mm caps in your favourite colour. Larger, 29mm caps are only found on Champagne-style bottles.

Use brown glass bottles. If a brewery cares about its beers, it will bottle them in good, brown bottles. Even green glass lets through loads of UV and risks skunking. Glass bottles are simple to sanitise and easy to cap. If you're lucky enough to get hold of (or be gifted) some swing-top brown bottles, feel very fortunate. But don't go out and buy them especially. They're expensive. Spend your money on good beer instead and save the empty bottles.

How to take a gravity reading

Cleaning & sanitising

1. Clean your bucket with hot water and a sponge. Rinse thoroughly

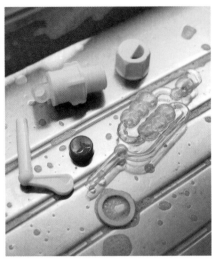

2. Disassemble your airlock and tap. Clean and rinse all individual bits

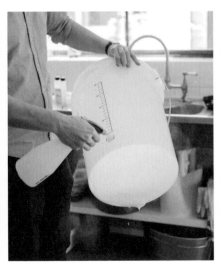

3. Sanitise your bucket's tap hole, inside and out

4. Sanitise each piece of your tap, and assemble within your bucket and then sanitise your entire bucket and lid

Brew pot

This is the first and most important piece of equipment to think about. Your choice will determine how much beer you can make and how strong it can be. It is the vessel in which all your water and grains will be steeped during the mash, then what your wort and hops will be boiled and cooled in.

Option 1: Stockpot (10+ litre/quart)

Downsizing to an 11-litre/quart stockpot was the best brewing investment I made. Rather than taking up the entire kitchen, a brew would take up only one hotplate on the hob. I found that by brewing really strong and then diluting (known as *liquoring back*), I could make nearly full-size batches anyway. Cleaning was a doddle. Everything happened so much faster.

If you are brewing 10-litre batches (just halve all the quantities in each of the recipes) then you can even get away without a wort chiller, too (see page 30).

Option 2: 30-litre/quart plastic boiler

This is the old-school, thrifty way of doing full-sized batches. Grab yourself an extra bucket and a kettle element; both available from all good home brew stores. You can then combine them to create a cheap boiler. Many home brew shops will do this for you, and you can buy them ready-made.

While this is a cheap way to reach higher capacities, and high-density polyethylene (HDPE) buckets can happily take heat, they do become flexible when full of boiling water. Try not to spill loads of boiling hot sugary stuff on yourself, or injure yourself in its production. They will also discolour with time.

Option 3: Big and shiny stainless steel pot

If you are going down the big and shiny route, I would recommend you plump for a pot that is **50 litres/quarts**, with this volume you can comfortably make 20-litre/quart batches of ludicrous strength, as well as 40-litre/quart batches of moderate-strength beer if you so wish. Most importantly, if you go for a 50-litre/quart pot then your buckets and all your other equipment can fit inside it for storage. Your entire home brew system sits within its footprint.

You'll still need a *heat source* – the options are between a kettle element or direct heat (gas or electric burner). If going for an element, have your supplier pre-drill a hole; high-quality stainless steel is a nightmare to drill through. For induction hobs, check with your supplier that the pot is magnetic and thus induction compatible. Gas rings are easiest – you can plop your pot straight on top.

You'll also want a *fitted tap* on your pot, for ease of transferring. I would go for a *stainless steel 1.5cm/½ inch BSP ball valve,* with a *hose barb* to go on the outside. On the inside, you want a *bazooka-style hop filter*. Anyone specialist enough to sell you a large pot should also be able to fit these for you.

Grain bag

When we brew, we need something to filter out all the spent grain after the mash. Traditionally, brewers mash in a mash tun, which has a mash filter. When we Brew In A Bag (BIAB), we use a grain bag.

You can buy a grain bag from any home brew shop – make sure you get one that is as big or bigger than your brew pot. If you are handy with a sewing machine, then you can make your own cheaply and easily out of a set of voile curtains and a nylon rope for a drawstring.

Insulation

The easiest way to keep your mash temperature constant is to wrap your big pot in duvets and blankets. Smother it. However, this doesn't leave you with easy access for checking the mash, stirring or checking the temperature.

By wrapping your pots in a form of insulation, you can avoid this faff. The most economical material is the foam, foil-backed **camping mat**, available from any large supermarket. Two more layers cut to the height of your pot and wrapped around, with another two layers on the lid, will be plenty of insulation. I've used duct tape to secure, and they've never come loose, even with regular drenching in both beer and water.

Wort chiller

Chilling your beer fast has many advantages – it reduces infection risk, it prevents cloudiness known as 'chill haze' and it saves you loads of time on brew day.

For smaller pots (<15 l/qt) that will fit inside a sink, a dedicated wort chiller is overkill. In this case, you can put the lid on, seal it with cling film and fill the sink with cold, preferably iced, water. Leave your pot to cool, replacing the sink-water as it warms. Swirling the pot will decrease cooling time markedly.

Wort chiller

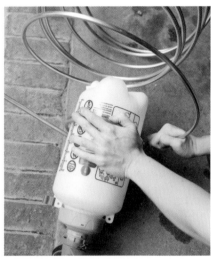

1. Spread out your copper coil, this makes it easier to work with

2. Bend it round something hard and round, taking care not to kink the tube

3. Keep bending the coil, pulling it tight, leaving about 1 metre (40 inches) at both ends to be safe

4. Bend these final lengths upwards so that they can hang over the edge of your pot, and tighten your hose over the end

For bigger pots, having a **copper immersion wort chiller** is recommended. You can pick them up ready-built from most home brew suppliers. If you're handy with compression fittings and don't mind a bit of DIY, you can make one yourself (see the previous page), but it is quite a bit of work bending the copper. There are plenty of guides online – it will save you a few quid.

Bag suspension

For full sized brews, holding the bag above your pot by hand to let the wort drain out becomes difficult, nay impossible. For these situations, I find it's best to rig yourself a wee suspension system. As long as you can find something that's going to be strong enough and high enough to swing a length of rope around and suspend a bag above a pot, you'll be golden. At home, I use a **stepladder and a few metres of nylon rope** (any sort of rope will do) – it works an absolute treat and is dead quick to set up. If this is an issue – don't worry. You'll need a spare plastic container (make sure it's food safe) or another large pot. When it's time to drain, simply remove your bag full of wet grain and stick it in here. You can also rinse it here with hot water to *sparge.*

Brewing software

OK, this bit's going to be a bit controversial. You've gone to the trouble of buying this book, and now I'm saying there's an app for that? Yup.

You'll thank me later. Brewing beer involves so many calculations, and brewing software can do it all for you. Yes, you can follow my recipes exactly, but you'll still get slightly different results than I did because there are so many variables to take into account. It helps to have something that keeps track of every variable you put in, and lets you know how these variables will change your beer. Take hops, for example. Every single year, each harvest is totally different. By inputting your specific hops' properties, the brewing software will tell you exactly how bitter your beer is going to be this year.

I recommend that you go for a single, *all in one* piece of software, and stick to it for good. Into these, you can input your entire recipe and it will throw every number you need right back at you. It will record all your recipes and the numbers you achieved, allowing you to compare, rank and potentially revisit every single beer you've ever brewed. I can't do that for you.

My choice is BeerSmith, a piece of software for PC or Mac (avoid the phone or tablet versions) that allows you to do pretty much everything. It's not the most intuitive, or indeed the best laid out, but it allows you to control pretty much every variable in brewing, and customise all your expected outcomes based on the results you've had from your own equipment. Then, it takes you through the entire brewing process step by step. It even sets timers for you. It's everything you need, but it isn't free. It's got a 3-week free trial, so you can see if you feel it's worth $27.95, or whatever they charge for it nowadays.

If that's not up your street, you could go for *BrewCipher*, a range of calculators for recipe creation integrated into a spreadsheet. This is completely free, provided you have software that can open and edit spreadsheets, such as Microsoft Excel. It has a great user guide and surprising functionality. The only thing you'll need to do is remember to save each recipe you create manually in another spreadsheet, as it will not collate.

A few free, web-based options exist. *The Grainfather Community* is probably the most professional, prettiest and simplest of these – for me, its single best function is the ability to share recipes with friends and fellow brewers online. It has a free version with everything a beginner might need. It is intuitive and easy to use.

Brewer's Friend is another online alternative that allows you to create recipes, use their individual brewing calculators or print 'brewsheets' on which to fill out your variables manually. You don't need to sign up, which is good. Brewer's Friend also has a free app for Apple and Android, which has a few basic calculators for things like alcohol by volume, hydrometer temperature adjustment and priming sugar. It's very handy, but there are loads of other apps that do exactly the same thing and this isn't a replacement for your primary piece of brewing software. Pick one of the above, and stick with it.

Carboys, demi-john & auto-syphons

I recommend sticking to using buckets as your primary and secondary fermenters. They're superb, despite the stigma – you can use a tap for taking samples and transferring beer, you can pop the lid off for easy cleaning and you can store them all inside each other.

But there are two disadvantages: buckets carry the risk of infection and they carry the risk of oxidation.

A **glass or PET (clear plastic) carboy** has a completely smooth inner surface, and a small opening on top. This means that bugs are less likely to fall in, less likely to grow, and you cannot scratch the inside through too-vigorous cleaning. PET and glass are nigh-on impermeable to oxygen; HDPE (bucket plastic) is not. If you're aging beers over a long time, you should take this into account.

Carboys do have drawbacks, and that's why I don't recommend them straight away. To take a gravity reading, for example, you'll need to suck a sample out, using a sanitised turkey baster or syphoning off a small amount using a silicone tube. These methods are annoying and, despite best precautions, risk infection.

Another drawback is that there isn't a tap. To transfer beer into your bottling bucket, you have to create a syphon. You can fill a length of tube with sanitised water, clipping the end and using the momentum to suck the beer through (never suck using your mouth) or use an **auto-syphon**, a nice wee piece of equipment that you pump down once to create the syphon effect. This does, however, have three separate parts and is quite difficult to sanitise confidently.

Because you can't get inside a carboy with your hands, they do require a different approach when cleaning. You have to think about it in advance, leaving them to soak in a cleaning solution such as Powdered Brewery Wash (PBW) or an oxygen-based solution like Vanish or Oxyclean.

Temperature control

I've said it probably ten times in this book, but I'll say it again: a good fermentation is the most important route to good beer. Good temperature control is an essential part of a good fermentation. If you want accurate and automatic temperature control, the easiest way to achieve this is to use a fermentation chamber. To you and I, that's an **old fridge** plugged into a **temperature controller** with two plug sockets on it. When the temperature probe inside the fridge reads too high, the socket into which the fridge is plugged triggers. If it reads too low, then it triggers a tube heater or heat bulb connected to the other socket, that sits inside your fridge. I recommend the Inkbird temperature controllers, and they're pretty cheap. Or you could just build one yourself. It is your responsibility to have your work checked by an electrician. Don't sue me, please.

All that's left to do is convince your cohabiters that you need another fridge dedicated to fermenting beer.

Ingredients

Grain

The *grain bill* provides the sugars that will eventually turn into beer. *Malted barley* makes up the bulk of it in most cases.

Malting is the process of soaking a grain in water. This allows it to germinate, producing enzymes that break down and release its reserves of starch. In nature, this allows the grain to use its stored energy to grow. In brewing, it allows us to utilise its energy to make beer.

Before a plant begins to grow from each grain, the process is arrested by quick drying using hot air from a kiln – a process known as *kilning*. Malts that are kilned for longer have fewer enzymes present, which means that they aren't so good at converting starches to sugar during the mash. That's why you'll always need a healthy chunk of pale malt in your recipes.

After drying, the final stage in malt production is *milling* – this is the process of crushing the grains into *grist*, to further release their starch. It is traditionally done just before brewing, as it's cheaper to buy unmilled grain. Until you're ready to buy grain by the 25kg/55lb sack, don't bother with a mill – just buy your grain pre-crushed. Use it within a year or so and it will work just fine.

Base malts

Base malts make up the bulk of our beer. Many of them are very similar, but they can be blended to add great complexity of flavour to any beer, regardless of colour.

Pale malts

For nearly all your beers, 'pale malts' will make up well over three-quarters of your grain bill. In the USA, the main options are **two-row**, for brewing lighter-bodied beers, and **six-row**. The latter has more enzymes and will give a more silky, bigger-bodied beer due to its higher protein content. Don't get hung up on their names – they just refer to the appearance of the barley's head.

In most of the rest of the world, we exclusively use two-row malts, but we have significant variation in how we produce them. Like which grape you choose decides your wine, which malt you choose decides your beer.

In the UK, *Maris Otter* and *Golden Promise* are great choices of pale base malt. They can add character to any British, Irish or American ale, and as such I use them in many of my recipes.

Lager malts

Pale malts made using even lower kilning temperatures. Lager malts contain higher levels of s-methyl methionine (SMM) than pale malts. SMM, during the mash, is broken down into another compound called dimethyl sulfide (DMS), which smells and tastes vegetal. DMS is easily taken care of by boiling your lagers for a longer time than you would a beer made with pale malt.

Lightly toasted malts

These are coloured malts, lightly roasted in a high-temperature kiln, but they still have enough enzymes to convert themselves. The most common are *Munich malt* and *Vienna malt*. Vienna is used to make full-bodied, biscuity Vienna-style lagers and bocks. Munich malt is slightly darker, and is used to add colour, nuttiness and mild sweetness to amber lagers, such as Märzens.

Wheat malt

Malted wheat has equal diastatic (enzymatic) power to malted barley, and its lack of an outer husk means it has fewer tannins. It has loads more protein, though, so in moderation gives a silky mouthfeel and better head retention. German wheat beers can be made with up to 100% wheat malt, but rarely do they contain any more than 70%.

Smoked malts

Instead of *kilning* your malt, you can dry it over an open flame. In Germany, beechwood is burned beneath the malt to give rauchmalz, a malt used to brew *Rauchbier*. Because of the relatively low temperatures involved, this malt is pale and can make up 100% of your grist.

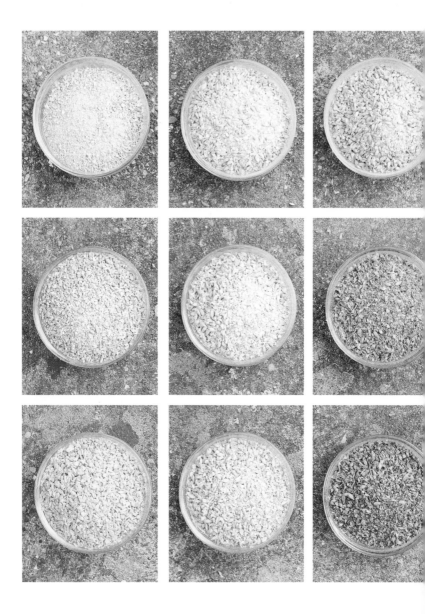

Malt colour chart

1	4	7	10	13
2	5	8	11	14
3	6	9	12	15

1   RYE MALT
2   WHEAT MALT
3   UNMALTED WHEAT
4   PILSNER
5   CARAPILS
6   MUNICH MALT
7   MARIS OTTER
8   PALE CRYSTAL
9   CRYSTAL
10 BROWN MALT
11 SPECIAL B
12 EXTRA DARK CRYSTAL
13 CHOCOLATE
14 ROASTED BARLEY
15 CARAFA 3

Speciality malts

These are malts that are added for flavour and colour. They do not have much enzymatic power. Steeping in hot water is usually sufficient to extract their desired properties, a technique utilised by those who brew using only malt extract. It's easiest and fastest for all-grain brewers simply to chuck them into the mash.

Crystal (caramel) malts

Crystal malts (caramel malts) start off with a soaking, just like any other malted barley. Then, instead of being dried in a kiln, they are heated gently. This activates their enzymes and converts their starch into sugars. Only then is the malt heated properly, which caramelises these new sugars as the malt dries. This heating results in the formation of dextrins – long-chained sugar molecules that are unfermentable by normal yeasts.

As such, they add sweetness and body to a beer, as well as a caramel-like flavour and colour. It is never wise to use crystal malts to make up much more than 10% of your grain bill. They vary in colour and the darkest malts will impart some roasted flavour.

Toasted malts

Malts that are roasted at high temperature are toasted malts. These tend to be powerfully flavoured and should be used in moderation. Overuse makes your beer taste like bitter burnt biscuits. Try not to go over 5% of your entire grain bill.

Amber malt (British) is a common toasted malt. It tends to be used lots in porters and brown ales, imparting a toasted bread flavour. A darker version is *brown malt*, which adds a mild coffee/cocoa flavour without the properly dark flavours of the blacker malts. *Biscuit malt* is a mellower and altogether more pleasant version from Belgium.

Roasted malts

These malts are produced in exactly the same way as toasted malts, but heated for longer and at higher temperatures. They appear black. They're used to make stouts and porters, adding strong flavours of chocolate and coffee. Roasted wheat, rye and spelt malts are all available. You'll usually see these marketed as chocolate malts, with pale chocolate as a slightly toned-down option. Use a combination and don't go above 10% of your total grain.

IBUs

Hops need to be boiled if they are to provide significant bitterness. Alpha acids are technical compounds found in hops that provide the bitterness. They themselves aren't bitter, but when they are isomerised at temperatures over 79°C (174°F), they become bitter and dissolve in the beer. The longer a hop is boiled, the more bitterness it gives. This bitterness is measured in IBUs – international bittering units. It's academic, but interesting: 1mg of isomerised alpha acid in 1 litre/quart of beer is 1 IBU. This is a tiny, tiny amount. The limit to human perception of bitterness is around 100 IBU – going above this will not make a beer taste more bitter.

Unmalted adjuncts

Wheat, rye, barley

Unmalted wheat is an addition that provides significant protein and complex carbohydrates to a beer. Rye and barley similarly so, but with more earthy and spicy flavours. In British and American beers, a small amount (<5% of grist) can aid body and head retention.

Oats

I bloody love oats. I add them to every beer I can get away with. They add body without adding any sweetness. A beer with lots of oats has a beautiful, silky mouthfeel, without being cloying. An oaty beer is drinkable, dry and mouth-coating. Oaty beers are awesome.

Roasted barley

This is an extremely dark, nay black, grain that is made by roasting unmalted barley until black. It was traditionally a primary ingredient in a stout, and provides a very harsh, burnt bitterness and some coffee and chocolate flavours. I feel it should be avoided, for the most part, and so you won't find it in my stout recipes.

Brewing sugars

Table sugar (sucrose)

Use it to add dryness and booze without affecting flavour, especially in IPAs and double IPAs. It pairs well with small amounts of unmalted oats and wheat, as these can prevent a dry beer from becoming thin. Don't go more than 10% of grist. Never add sugar expecting sweetness – the yeast will take care of that.

Belgian candi sugar

Belgian candi sugars are actually syrups of varying darkness, made up of sucrose that has been heated until it forms a caramel. The darker a syrup you choose, the more of these burnt-sugar flavours you're going to get. In beers such as Belgian strong dark ales, these contribute a raisin-like, dark fruit flavour, alongside the sugar's inevitable dryness.

Honey

To get a really good honey flavour, you've got to use really good honey. This is expensive. Treat it delicately, adding it after primary fermentation, so that as few of the aromatics as possible are boiled off.

Lactose

Most commonly used in milk (sweet) stouts and to back-sweeten ciders that have gone too dry, lactose cannot be metabolised by normal brewing yeast and therefore persists, unlike table sugar. Except in sour or wild beers, where the Brett (see page 65) will take care of it.

Hops

Hops, as far as we're concerned, are the female flowers of the hop vine, *Humulus lupulus*. The flowers look like little green pine cones, and contain essential oils that we can utilise for flavour and bitterness. They are picked, dried and then packed into sealed containers to keep fresh.

Beer would not be what it is without hops. Some go on about their antimicrobial, preserving qualities, but this isn't why hops are used in beer today. Above all, hops provide balance. Their bitterness restrains the sweetness from the malts, increasing drinkability. Their aroma adds complexity to the beer, preventing it from being one-dimensional. They make beer interesting. They make beer beer.

Bitterness vs aroma

Hops can be crudely divided into 'bittering' and 'aroma' hops. Bittering hops tend to have a higher '*alpha acid*' content than aroma hops, and so provide more bitterness when added during the boil. Aroma hops tend to be added near the end of the boil so as not to boil off their complex aromatics.

This is a traditional division. Most modern hops are dual purpose – many bittering hops give a really interesting aroma, especially when balanced with other hops. Aroma hops are perfectly capable of providing bitterness, too; you just need more of them and hops are expensive.

The *alpha acid percentage (%AA)* indicates the bittering potential of your hops. It varies wildly between harvests and between hop varieties. Boiling a hop with a higher alpha acid percentage will make your beer more bitter. These higher-percentage AA hops tend to be those that are traditionally classed as bittering hops.

When to add hops

Bittering additions
These are hop additions added at the beginning of the boil. Over the next 60–90 minutes, their alpha acids are *isomerised*, creating bitterness. Any aromatics are completely boiled away, leaving no aroma.

First wort hopping (FWH)
This is a slight variant on the above and using it is a practice I recommend heartily. It involves adding your hop additions as soon as the lauter is over. Theoretically, this allows hop oils to oxidise and therefore become more soluble as the beer heats up, helping their flavours survive the boil.

'Flavour' additions
During the final 30 minutes of the boil, hops can be added to give a compromise between bitterness and aroma. During this time, not all of their aromatic compounds will be boiled off, and therefore they will impart some character. Adding lots of aroma hops in the last 15 minutes, foregoing any bittering additions at all, is called *hop bursting.*

Aroma steep
After the boil has finished, many brewers add more hops as a 'flameout' or 'whirlpool' addition, then allow these hops to steep for between 10–30 minutes. I recommend you chill the wort to below the threshold of isomerisation (79°C/174°F), so you can add as many hops as you like, and steep them for as long as you like, without imparting more bitterness.

Dry hop
This is when you add hops to your fermented beer, It's best done at room temperature for a short period of time, as this can prevent the 'grassy' flavour that's traditionally associated with dry hopping. I'd go for 3 days, at 18–21°C (64–70°F).

Pellets vs Whole Leaf:
what to buy and how to store

Whichever type of hop you go for, freshness is key. Hops should be bought in a nitrogen-purged vacuum-packed container. They are best stored unopened at as low a temperature as you can get – the freezer. Once opened, you should wrap your pack in cling film and freeze.

Always judge if a hop is fresh by its smell – if a hop smells cheesy or even underwhelming in any way, don't use it and ask for your money back. Fresh hops should blow your mind.

Hop pellets are finely ground-up hop flowers. They dissolve easily in your beer and are an excellent choice for dry hopping. The only issue with them is you can't use them with traditional hop filters, so you'll need to use a hop bag to stop them getting everywhere. But they will settle to the bottom at the end of the boil, or if you chill your fermented beer.

Whole leaf hops are precisely what they sound like. They're simply the whole female hop cones. The difference in flavour between whole leaf and pellets is minimal, but whole leaf tends to absorb far more of your beer or wort, leaving you with less at the end. If you're using a bag anyway, go for pellets.

Common hop types

There are loads and loads of hops out there. Which you choose to use is completely up to you, but these are a select few of the varieties to look out for. I can personally advocate for the brilliance of all of them.

American hops

Origin/variety	Bittering or aroma	Alpha acid %	Flavour	Substitute
Amarillo	Dual purpose	8–11%	Distinct orange, tangerine and grapefruit aromas	Centennial, Citra, Mosaic
Apollo	Bittering	15–19%	Clean bitterness, dank if used for flavour or aroma	CTZ, Warrior, Magnum
Cascade	Dual purpose	4.5–7%	Pine and citrus; mild by US standards	Centennial
Centennial	Dual purpose	9–12%	Cascade on steroids; intense floral and citrus notes	Cascade, Amarillo, Chinook
Chinook	Dual purpose	12–14%	Good bittering qualities; resin, pine and spicy aroma	Centennial, CTZ
Citra	Dual purpose	11–13%	Intense citrus, gooseberry, passion fruit, lychee. Also excellent bittering qualities	Amarillo, Mosaic, Centennial
CTZ (Columbus, Tomahawk & Zeus)	Dual purpose	14–18%	Dank if used as aroma hop; excellent bitterness	Chinook, Centennial
Magnum	Bittering	12–14%	Exceptionally clean bitterness	CTZ, Warrior
Mosaic	Aroma	11.5–13.5%	Tropical fruit, intense citrus and red berries	Citra, Amarillo, Galaxy

American hops (continued)

Origin/variety	Bittering or aroma	Alpha acid %	Flavour	Substitute
Simcoe	Dual purpose	12–14%	Distinct passion fruit, pine and pleasant cat-pee	Summit, Chinook
Summit	Bittering	16–19%	Spicy, earthy. Good bitterness	CTZ, Warrior
Warrior	Bittering	15–17%	Mild, resinous. Extremely soft, rounded and clean bitterness	CTZ, Magnum, Summit

English hops

Origin/variety	Bittering or aroma	Alpha acid %	Flavour	Substitute
Challenger	Dual purpose	6.5–9%	Wood, green tea	Any English hop
East Kent Goldings	Dual purpose	4–6%	Mildly floral, spicy and earthy. Best choice for any English ale	Any English hop
Fuggles	Aroma	3–7%	Mild sadness; old men; grass	Anything else
Northern Brewer	Dual purpose	8–10%	Pine character	Chinook, East Kent Goldings, Target
Target	Bittering	8.5–13.5%	Good bittering characteristics; spicy and sage aroma	Magnum, Any English hop

European hops

Origin/variety	Bittering or aroma	Alpha acid %	Flavour	Substitute
Hallertau Mittelfrüh	Aroma	3–5%	Mild, floral and spicy	Saaz, Tettnang
Hersbrucker	Aroma	2–5%	Mild herbal, medicinal and fruity	Spalt, Tettnang
Perle	Aroma	6–10%	Mild fruit, mint and spice	Northern Brewer
Saaz	Aroma	2–5%	Very mild. Floral and spicy	Hallertau, Tettnang
Spalt	Aroma	2–6%	Mild, spicy and floral	Saaz, Tettnang
Tettnang	Aroma	3–6%	Earthy and herbal; mild	Saaz, Hersbrucker

New World hops

Origin/variety	Bittering or aroma	Alpha acid %	Flavour	Substitute
Galaxy	Aroma	11–16%	Citrus fruits, peach. Very similar to *Mosaic*	Citra, Mosaic, Amarillo
Motueka	Aroma	6–9%	Strong tropical fruit, lemon and lime, orange blossom	Amarillo, Mosaic, Nelson Sauvin
Nelson Sauvin	Dual purpose	12–13%	Lychee, gooseberry, Marlborough Sauvignon Blanc	Mosaic, Citra

Water & water treatment

It's important not to forget about the ingredient that makes up more of your beer than any other – water.

Before you brew using the mains supply, you should ask yourself: do you like the taste of your water? If the answer is hell no, it's probably indicative that you might want to treat your water in some way. Water can taste bad because it has too much minerality, but also if it is lacking in it.

Treating water is a doddle. Find your water report for your local area (contact your water supplier – often it's online) then it's just a case of adding a pinch of powder at the start of each brew. The compounds to really pay attention to are **bicarbonate (HCO_3^{-1})**, and **calcium (Ca^{+2})**.

Bicarbonate (HCO_3^{-1} – total alkalinity)

Treatment required: if above 150ppm (parts per million) for pale beers; if below 100ppm for dark beers

Bicarbonate is what you want to look for first on your water report. You'll know if your bicarbonate is high, because you'll know if you have hard water. If you boil hard water, *chalk* (calcium carbonate) will precipitate and leave you with a white mess on your kettle and your brewing equipment. Because of its alkaline qualities, bicarbonate has an important impact on the pH of your mash. Good control of pH is crucial for an effective mash and therefore good conversion of starch to sugar.

Treatment:
For most people with a little high bicarbonate, ½ teaspoon of *gypsum* (calcium sulphate) goes a long way towards getting the right level of acidity. At high levels over 400ppm, you can dilute with shop-bought water or you can pre-boil to precipitate your carbonates as chalk first.

Calcium (Ca^{+2})

Treatment required: if below 50ppm

Although calcium partly determines your water hardness (along with bicarbonate), calcium is essential for the correct functioning of several mash enzymes – it is a yeast nutrient and ultimately gives clarity and stability to your finished beer.

Treatment:
It's unlikely your calcium levels will be too high. If they're low, add ½ teaspoon of *gypsum*. For most of us, it's never that bad an idea to add a little gypsum.

Others to consider:

Sodium (Na^{+1}) levels add saltiness to your beer. In mild levels, this simply gives fuller mouthfeel and acts as a flavour carrier. As soon as it becomes detectable, it's just a bit weird. It's like drinking seawater.

Sulfate (SO$_4^{-2}$) levels add a drying effect to beer, which can accentuate hop character. Add ½ teaspoon of gypsum in pale hoppy beers *after* the mash, even if your water requires no treatment.

Chloride (Cl^{-1}) can enhance mouthfeel in low concentrations, but above about 300ppm it can impart a harsh, medicinal taste not unlike TCP. You can treat using potassium metabisulphate, which you can get from all home brew shops.

Water sterility

On your water report, there will be a bit about the number of bacteria, such as coliforms, that are present in your water supply. This information is useful. If the number of these bacteria (noted as 'CFUs') in your water report is 0 or anywhere in the single figures, you know that your water is pretty much sterile. Much more sterile than bottled water, even.

To infect a beer, you need a significant dose of a beer pathogen (bacteria or wild yeast). If your tap water contains even small amounts of bacteria, the likelihood of getting an infection from it is negligible. Do not worry about liquoring back (diluting) to hit your numbers and making yeast starters with tap water – so long as you sanitise the tap first, you'll be fine.

Yeast

In the introduction, we have already talked a little about yeast, and some of the different characteristics you might expect from its various strains. This is a simple enough guide to get you through, if you want to stick to it. But I implore you to delve into the depths of yeast. It's fun, honest. However, yeast is something that you could write an entire book on. In fact, several books exist on the subject. *Yeast: The Practical Guide to Beer Fermentation* by Chris White and Jamil Zainasheff is a good one to check out.

I will reiterate: fermentation is the most important phase of the brewing process. The fermentation will decide, above all else, whether your beer is going to be good or not. Having a good fermentation requires that, first and foremost, good sanitation practices are followed. Then you need to **understand your yeast**. You need to have **enough yeast**. You need to have the **right yeast**.

Understand your yeast

Brewer's yeast, or *Saccharomyces cerevisiae*, is a single-celled fungus. Its job, from our biased perspective, is to sugars into ethanol. Handy by-products of this process include CO_2, which makes beer fizzy, and various delicious flavour compounds.

Oxygen is required for yeast to synthesise new cell membranes from fatty acids, and to synthesise the molecules that allow it to absorb maltose. Maltose makes up most of the sugar it will be absorbing during fermentation of beer.

Because of this, when you want yeast to grow, you should give it oxygen. In practical terms, when you're making a yeast starter you need to make sure it is well aerated. The easiest way for a brewer to do this is to shake their starter, vigorously and regularly.

Another factor to take into account is temperature. Yeast will grow in conditions up to 37°C (98°F), or body temperature, but at these high temperatures they will be seriously stressed.

Important yeast terms

Flocculation

'Flocculation' is when yeast stick together after fermentation is finished, causing them to sink to the bottom of your fermenter. High levels of flocculation are seen as a positive characteristic in a yeast strain, because this leads to clearer beer. However, if yeast flocculate too early, this will lead to sweet, under-attenuated beer.

Wild yeasts, such as those that can infect a beer, do not flocculate except with very cold temperatures. As such, they can remain in solution and have free reign over a beer, as the yeast strain you intended to ferment the beer sits at the bottom.

Attenuation

'Attenuation' is the degree to which the sugars in a beer are metabolised by the yeast. The most important factors that influence attenuation are temperature and yeast strain. If a beer is fermented too cold, for example, the yeast will flocculate too early and leave you with a sweet, cloying liquid. This is an under-attenuated beer.

In beers that we want to be very dry, such as Saisons or double IPAs, we want very high levels of attenuation. To achieve this, we can use a yeast strain that's known to attenuate highly.

Each yeast strain has a known range for its attenuation and you can use this to work out if your beer is likely to be finished fermenting or not. For example, White Labs WLP001 has an expected attenuation of 73–80%. This means that 73–80% of the sugars appear to have been metabolised, as measured by a drop in specific gravity.

It's easiest to use your brewing software or an app to work it all out, but if you want you can work out your beer's % attenuation using the formula below. Then, you can cross reference this with the expected range for your yeast.

% attenuation = [(original gravity – final gravity) / (original gravity – 1)] x 100

Instead, you should attempt to keep your yeast fermenting around room temperature, or 18–21°C (64–70°F). Be wary, though, as yeast produce heat when they ferment, and so a fermenter can be several degrees higher than ambient temperature during active fermentation.

Yeast will only grow, too, if they have the right nutrients. Thankfully, you don't need to go out and buy that 'yeast nutrient' you might find for sale in your local home brew shop. Wort contains pretty much everything a burgeoning yeast culture might need – amino acids, fatty acids and vitamins. This is, in part, why you shouldn't try to grow yeast using only a sugar solution.

Pitch enough yeast

Yeast is available in two main forms – dried and liquid.

Dried yeast tends to come in packets of 11g, which equates to around 200 billion cells in one pack. These can be sprinkled straight onto your cooled wort, but this is probably not wise. The change in environment can shock the yeast, leading to a slow lag time and significant cell death. To prevent this, dried yeast should be rehydrated before pitching.

Dried yeast cell counts do fall over time, but very slowly. If you keep a packet of dried yeast in the fridge, you've got years to use it.

Liquid yeast comes in vials or packets, and usually contains 100 billion cells at the time of packaging. Their disadvantage is that, when kept in the fridge, more than 20% of these cells will die every month. This means that their useful shelf life is practically 2 or 3 months.

Liquid yeast vials rarely contain enough cells to ferment a standard-sized batch of beer on their own – you are expected to use several, or make a *starter*. A starter is a small volume of wort that you make using dried malt extract (DME), and ferment it using your yeast vial. This causes the yeast to multiply, leaving you with enough yeast to ferment your beer.

You should never make a yeast starter from packets of dried yeast.

The dried granules contain not just yeast, but enzymes and nutrients that help the yeast in their initial fermentation. If you make a starter, these are used up and so aren't utilised during fermentation of the main batch of beer. By attempting to make a starter in this way, you can actually end up with even fewer cells than you started with.

Having enough yeast is extremely important. The number of yeast cells you add to your beer is called the *pitching rate*, and it is usually referred to in cells per millilitre. Your yeast requirement will depend on how much sugar the yeast has to get through: how strong your beer is. For a very weak beer, pitching rates start at around 6 million cells per ml (yup, the cells are that small) and go up to 20 million cells per ml (or more) for very strong beers.

Practically, we need to know how many packets or vials to add. Thankfully, there are calculators for that – doing this by hand would be impractical. Search the internet for a 'yeast starter calculator' or use your brewing software's in-built functionality

Once you've calculated the number of yeast cells you need, you'll have a number. You won't hit that number exactly – you'll always either *over-pitch* or *under-pitch*.

Adding too little yeast causes rapid growth during the first couple of days of fermentation and so loads of off-flavours can be produced. Then, despite this growth, your overall yeast count is still going to be low. Your yeast will be stressed and struggle to cope with their burden, leading to a lacklustre fermentation. This causes them to flocculate fast and under-attenuate.

Pitching *too much* yeast gives you a short lag-time with less growth, leading to limited flavour production so you end up with a one-dimensional flavour profile. Your beer will have a vigorous fermentation, giving an over-attenuation and a very dry finish. If you were after a beer with body, you might find yours is a bit thin and strong.

It's best to pitch the right amount, but **if you don't pitch enough yeast, you'll have more problems than if you pitch too much**.

Pitch the right yeast

You can get a healthy fermentation with any yeast, but a healthy fermentation with the wrong yeast will lead to a strange beer. Your yeast choice will depend principally on the type of beer you want to make. In my recipes, I've recommended specific strains or brands, but the range is developing all the time and thus you might want to try out something different. Here are the loose categories and some good examples:

American ale yeasts
American ale yeasts tend to be low-medium flocculators (leaves a moderately cloudy beer) and give high attenuation, leaving a very dry, clean and crisp finish to your beers. They complement hop character well.

English ale yeasts
English yeasts tend be highly flocculent (leaves a clear beer), low attenuators. This leaves residual sweetness, and can accentuate malt character. Because these yeasts fall to the bottom like a stone when they're finished, they tend to drag hop character and bitterness down with them.

Belgian ale yeasts

Belgian yeasts have a fruity, estery profile. You might notice they make a beer 'spicy'. They have low flocculation and extremely good attenuation, leaving very dry beer if used correctly. They might produce banana flavours if mistreated. You will want to keep them at the cool end of their temperature range for the first few days, as this limits their rate of growth and their production of off flavours. After two or three days, you can let them fly, and they will give you magical attenuation.

Hefeweizen yeasts

Wheat beer yeasts are some of the least flocculent around – they cause a perpetually cloudy beer. They have a specific ester profile, and if pitched right they will produce a balanced aroma of banana and clove. If you add too much of this yeast, you will get more banana flavour.

Hybrid yeasts

Hybrid styles include *California Common*, *Kolsch* and *Altbier*. These are ale yeasts of high attenuation and low flocculation that give lager-like results at normal ale temperatures.

Lager yeasts

Lager yeast is a different beast entirely – at anything above 15°C (59°F), most will produce pretty unpleasant tastes. Unless you have a fermentation chamber or other automatic temperature control, this is a faff. Stick to hybrid styles initially.

Kveik yeasts

This is a family of very resilient Scandinavian farmhouse yeasts. While they are of the *Saccharomyces* species like the other yeasts here, they behave rather oddly. They are extremely heat tolerant, producing few of the fusel (harsh, higher) alcohols that would be expected at temperatures of up to 35°C (95°F) or even higher. You can pitch tiny amounts, and end up with a relatively mild flavour profile.

Check out the descriptors and the extremely helpful article found at *milkthefunk.com* – just search for 'milk the funk kveik'.

Wild yeasts & bacteria

You can now purchase an ever-expanding range of 'wild' yeast and bacteria, in order to replicate sour or funky beers from across the world. Handily, the same bugs can cause infections in your beer, so you can understand why people don't like using them.

Brettanomyces sp.
Brett has a reputation as the worst bug that can infect a beer. It is extremely slow growing compared to *Saccharomyces*, and does not flocculate well. Many species of Brett can ferment nearly every source of energy left in a beer after brewer's yeast has done its job – including lactose, longer-chained sugars (dextrins) and starches. As a result, it leaves a beer cloudy, exceedingly thin and causes bottles to explode. It can take a long time to appear.

Brett can destroy or enhance hop character, and munch through the flavour compounds that make a certain beer what it is. It can produce some acetic acid, giving a sour twang. However, Brett can add astounding complexity. Bretted saisons are just better, I believe, than their meek equivalents. And if any beer has been barrel-aged in anything but casks straight from a distillery, the wood is going to have Brett burrowed into it. Brett makes things taste nice.

Pediococcus
Pedio is an acid-producing bacteria that gives a distinctive sour kick – it is responsible for the majority of acidity in traditional Lambics. *Pediococcus* can cause a beer to become 'ropey'. This is when you start to notice strings of horrible slime throughout your beer. (the common word 'ropey' came from beer infections). This slime is made of polysaccharides, which are unfermentable by normal yeasts. But Brett will munch right through them. Always pitch *Pediococcus* with *Brettanomyces*.

Lactobacillus

Another bacteria and another sour one – this is the principal bacteria you'll find in sourdough starters for making bread. It gives a very clean sourness, and like *Pediococcus*, it's quite hard to infect your beer with it.

Acetobacter

This is mostly bad. *Acetobacter* makes acetic acid. It turns your beer into vinegar.

I would never intentionally pitch *Acetobacter*. These infections tend to occur over a long time; you must already have a bacterial infection and you must have oxygen in your beer. This is unlikely unless you transfer or shake your beer regularly, or store it exposed to the elements. An *Acetobacter* infection is an indicator that you need to totally reassess your attitude towards sanitation, as well as keeping oxygen out of your beer.

Other ingredients

Finings

Even though we're home brewers, most of us still seek to create something that resembles a professional product. While it might not make much in the way of difference to the flavour of our beer, a beer that is clearer might just be perceived more positively.

To add clarity to any beer, you can add finings. There are two broad categories – those added during boil, and those added after fermentation.

Kettle finings are derived from an algae called Irish moss. You can buy this dried in its original form, or in tablets – Whirlfloc and Protofloc are the most common brands. Adding the finings 15 minutes before the end of the boil almost eliminates protein haze in a beer – big clumps settle to the bottom as trub, instead.

Additive finings, added after fermentation, will cause even more coagulation of proteins, polyphenols (such as tannins) and yeast. *Isinglass* is the most commonly used fining, especially in cask beer. It is derived from fish swim bladders, and this is the reason many beers are not suitable for vegetarians. *Leaf gelatine*, mixed with just-boiled water before being added to your fermenter, will remove both proteins and polyphenols. It is my first choice if I want a clear beer after fermentation and conditioning is complete.

Fruits

While it might be tempting to add fruits and juices to obtain their flavour, it isn't as simple as that. The flavours of most fruits change when their sugar is removed, and the yeast have had a go at them.

If you would like to add fruit juice, add it at the start of primary fermentation. No more than 1 litre/quart for every 10 litres/quarts of beer, I'd suggest. I can't vouch for what effect it will have on the beer.

Aging on whole fruit, on the other hand, can be wonderful in sour beers. Cherries, blueberries, apricots and raspberries are my favourite, pulped or whole (pulped means they ferment a bit quicker). Add them once almost all yeast activity has settled down – because they contain sugar, they'll stir things up a bit. Leave for at least a couple of weeks before transferring and bottling.

Brew day from scratch

Step 1: prepare the yeast

What you actually need to do:
- Work out how much yeast you need
- Be sanitary
- If using dried yeast, ensure sufficient quantity and rehydrate
- If using liquid yeast, make a yeast starter at least 2 days before
- If re-pitching yeast from trub, ensure sufficient healthy quantity

Using dried yeast

1. Work out how much yeast you need. Use an **online yeast calculator** or your chosen beer software to work out how much you need. Assume that each 11g packet has approximately 200 billion viable cells. Work out how much yeast you need to the nearest half-packet.

For example, if you were brewing 23 litres/quarts of beer with an OG (original gravity) of 1.060, you would need roughly 255 billion yeast cells to get your desired attenuation and a clean fermentation. Rounded up, this is 300 billion cells. Therefore, you should use one and a half packets of dried yeast in this beer.

2. Sanitise a small vessel, such as a jar or a pint glass. Boil some water, and half-fill the glass. Cover with sanitised cling film, and leave this to cool until it reaches 30°C (86°F). This can be done as you prepare your mash water.

3. Sanitise your packet of dried yeast, and sanitise the scissors or knife you use to open each packet. Pour your required amount of yeast into your warm water. Use your sanitised thermometer to stir.

4. Re-sanitise your cling film and replace it. Leave your yeast to rehydrate for at least 30 minutes, or up to several hours, before use (set it somewhere out of the way where it isn't going to be knocked over). When it is time to pitch the yeast at the end of your brew day, you should add all of this liquid.

Using liquid yeast

If you use liquid yeast, you've got to **make sure you have enough vials** (expensive), so that you have enough yeast cells to ferment your beer. Alternatively, begin to prepare a yeast starter at least 2 days in advance of your brew (see overleaf and page 31).

Liquid yeast vials contain around 100 billion cells at the time of packaging, but this quickly reduces over time. Use an online yeast calculator (such as yeastcalculator.com) or your brewing software to work out the number of cells left in your packet by entering its production date, which should be listed on the side. Then, enter your planned beer's OG and batch size into the same tool. This will tell you how many yeast cells, usually in billions, are required to ferment your batch.

Making a yeast starter is the simplest and cheapest way to increase your cell count. Below is a table showing how many billions of cells you'll get from your own starter size and cell count/age of your yeast vial. The starter OG is 1.040, meaning I use 100g/3½oz of dried malt extract for every litre/quart of water. I do this because it's easy to remember, and it's not so strong that the yeast is stressed out.

Option 3: pitching from trub

Vial cell count/age / Starter size	90 billion 11 days old	75 billion 1 month old	50 billion 2 months old	25 billion 3.5 months old	10 billion 4 months old
1 lt/qt	188	167	129	84	49
2 lt/qt	241	216	168	111	66
3 lt/qt	282	253	197	131	71
4 lt/qt	315	283	222	147	71
5 lt/qt	344	309	243	174	71

Making a yeast starter

1	4	7	10	13
2	5	8	11	14
3	6	9	12	15

1   SANITISE THE YEAST VIAL AND OPEN IT COLD TO PREVENT FOAMING

2   LEAVE YOUR VIAL WITH THE LID LOOSELY ON TO COME UP TO ROOM TEMPERATURE

3   SANITISE YOUR STARTER CONTAINER

4   SHAKE TO FULLY COAT WITH SANITISER

5   SPRAY SOME FOIL WITH SANITISER AND COVER

6   WEIGH OUT YOUR DRIED MALT EXTRACT (DME) IN A PAN – 100G/3½OZ OF LIGHT DRIED MALT EXTRACT PER LITRE/QUART OF WATER

7   ADD THE REQUIRED WATER TO YOUR DME AND MIX

8   BRING YOUR MIXTURE TO A BOIL IN A SAUCEPAN

9   REPLACE THE LID AND WRAP YOUR SAUCEPAN TIGHTLY IN CLING FILM

10 COOL YOUR WORT BY PLACING IT IN A SINK FULL OF COLD WATER. SWIRL TO COOL. WHEN COOL TO TOUCH, CHECK THE TEMPERATURE WITH A SANITISED THERMOMETER. YOU WANT BETWEEN 15 AND 25°C (60 AND 77°F).

11 POUR THROUGH A SANITISED FUNNEL INTO A SANITISED CONTAINER

12 OPTIONAL: YOU CAN SAVE TIME BY TOPPING UP A STRONG STARTER WITH TAP WATER FROM A SANITISED TAP

13 ADD YOUR YEAST VIAL

14 SWIRL TO AERATE

15 LEAVE YOUR STARTER TO FERMENT, SHAKING IT AS OFTEN AS YOU REMEMBER. AFTER 24 HOURS, OPTIONALLY STICK IT IN THE FRIDGE FOR 24 HOURS SO YOU CAN POUR AWAY THE CLEAR BEER ON TOP, LEAVING ONLY YEAST.

Making a yeast starter is actually just making a very small batch of beer. It stands to reason, then, that if you make a beer, you're just creating a very large yeast starter, right?

When you transfer a beer to the secondary fermenter or bottling bucket, you leave behind the layer of disgusting-looking trub, which is absolutely packed with healthy yeast. You can use a yeast calculator, entering in the original gravity (OG) of your beer and its volume, as well as how much you pitched, to work out roughly how much yeast is in there. Be aware that if your beer was higher gravity (over 1.070), your yeast will have **undergone a stress-filled fermentation and will not be as healthy**. Use yeast from strong beers at your own risk.

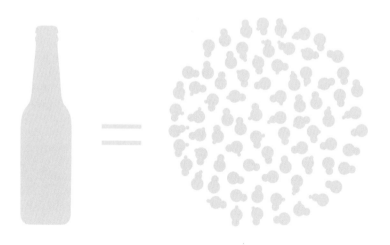

Step 2: water heating & treatment

What you actually need to do:
- Measure out how much water you need, from your recipe
- Use hot tap water if you have a combi boiler
- Turn your heat source on
- Add treatment to your water
- Wait for your water to rise to the right temperature

As covered in the previous chapter, your water is an important ingredient (see page 55). Think about whether it needs treatment. You're going to need to know how much water you'll need to brew your beer. If you're using one of my recipes, I'll tell you exactly.

If you lose too much water during the boil, and your beer is too strong, you can always add a bit more at the end. This is called *liquoring back* and it is the best way to be sure that you're going to hit your target numbers. If you've added too much water, the only way to remove it is by boiling more. This is a slow process and can make your beer extra-bitter, so it pays to brew a little bit on the strong side.

As a rule, **your weight ratio of water to grain during the mash should be at least 2:1**. For example, if you were making a beer that required 2kg/4½lb of grain, you must mash with, at the *very* least, 4 litres/quarts of water. However, you should always aim for a looser mash if possible. **Mash ratios of 4:1 or 5:1** are more appropriate for BIAB (brew-in-a-bag). Measure out how much water you need, as specified in your recipe. To save time, use water from the hot tap if you have a combi boiler. If you have a hot water tank, this can be full of mineral precipitants so use cold water instead.

Turn on your element and let the water heat up. Check it regularly.

Add your water treatment: see the previous chapter.

See your recipe for the exact temperature you want – this will be above your expected mash temperature. At this point, turn off the heat and you are ready to mash in (see the next step).

Brewing strong: liquoring back

There are a couple of reasons why you might go for a very thick mash. The first is simple – to make a very strong beer. More grain in a smaller volume of water means there is a higher concentration of fermentable sugar. The second is so far underexplored in brewing – so you can make a large batch of a regular-strength beer, using very small equipment.

Using the technique of brewing a very strong, bitter beer, I can happily make 15–20-litre/quart batches in a 10-litre/quart pot. A thick mash, at a water-to-grain ratio of 2:1, takes up 2.8 litres/3 quarts of space for every 1kg/2lb of grain. Therefore, I can easily make a beer using 3.5kg/8lb of grain in a 10-litre/quart pot. At my usual efficiency, this would give me an OG of 1.036 at 20 litres/quarts, or 1.048 at 15 litres/quarts. All I need to do is dilute (liquor back) my strong beer to hit my required volume.

It's worth considering that your choice of mash thickness will impact on your eventual beer. If you choose a thick mash, your mash's efficiency tends to be a little higher. However, the fermentability of the resulting wort is actually lower, as longer-chained sugars tend to be formed. This can lead to a sweet, more full-bodied beer. Furthermore, if you mash thick, then you should watch that you do not sparge with too much water, as this can lead to tannic, astringent beer.

Step 3: the mash

What you actually need to do:
- Weigh and mix all of your grains together
- Add your grains, mixing in well
- Check and adjust your temperature
- Leave your grains to convert for 1 hour and check on them regularly

The *mash* is when you mix your grains and your water together and then hold them between 64 and 69°C (147 and 156°F). Usually, you do this for about an hour, to allow your grains to hydrate and the certain enzymes to split the starches in the grains into sugars.

The enzymes are called amylases. *Alpha-amylase* is most active between 67 and 75°C (152 and 167°F), and can snap a starch molecule from anywhere along its chain. This results in a high number of long-chain sugars, called *dextrins*. These are unfermentable and they therefore add sweetness and body to your beer.

Beta-amylase is active between 54 and 65°C (129 and 149°F), especially at the top end of this range. It can only chop sugars, one by one, off each end of the starch molecules. Nearly every sugar that results is *maltose,* a fermentable monosaccharide. This leads to a light-bodied and dry beer.

Mashing for light body – 64–65°C (147–149°F)
If you want your beer to be as dry as possible, with high drinkability, you should mash at a lower temperature. You'll probably want to mash like this for most beers you make, causing near exclusive activation of beta-amylase.

Mashing for full body – 66–67°C (151–152°F)
At these temperatures, we're at a compromise of alpha- and beta-amylase activation, and thus we will have some dextrin formation. You'll want to delve into this range for the traditional English styles.

Mashing

1. Weigh out all your different grains into one container

2. Slowly pour your grains into your bagged boiler, stirring all the time

3. Leave the temperature to even out

4. After 5 minutes, check the temperature is within the range you're after. Adjust with boiling or cold water if need be

Mashing for very full body – 68–69°C (154–156°F)

At this point, there's a lot of alpha-amylase conversion going on and you'll have a very full-bodied beer resulting from a mash at this temperature.

Mash efficiency

There's a certain amount of starch in any given quantity of unmashed grain. If you were to extract every last morsel of this and convert it into sugar, you would have 100% mash efficiency. Unfortunately, most mashes are not that efficient. With brew-in-a-bag, we might aim for about 70–75% efficiency.

How to mash

1. Weigh out all your grains into a bucket. Mix them together to combine.

2. Place your grain bag inside your pot, securing it in place around the edge. Pour the dried grains into the bag, a little at a time, and stir as you do so.

3. Stir your mash thoroughly to remove the inevitable lumps. Place the lid on, set a timer for 60 minutes (unless otherwise specified) and leave it to sit for 5 minutes.

4. Check the temperature of your mash after 5 minutes (checking it straight away is not accurate). You want it within half a degree Celsius of your target temperature.

5. If your mash is too hot, you need to cool it down by adding 500ml/1 pint cold water, then stir. This should drop it by a degree. If your mash is too cool, heat it up by pouring in (roughly) 500ml/1 pint boiling water – this should increase it by about a degree. Stir, wait 5 minutes and re-check.

6. If you do not yet know your equipment, check the temperature of your mash at 15, 30 and 45 minutes. This allows you to heat it up using hot water if it has cooled. If you know your pot is well insulated, you might be able to leave it the whole hour without any significant drop.

Step 4: lauter

What you actually need to do:
- Raise your mash temperature to 75°C (167°F) ('mash out')
- Hold the grain at 75°C (167°F) for 10 minutes
- Suspend your bag above your wort
- Sparge 75°C (167°F) water through your bag to rinse your grains

By the time you reach the end of the mash, all the starches you're realistically going to convert to sugar will have been converted. The next step is to remove the grain from the liquid, leaving your hot, sweet liquid that is now known as 'wort' (pronounced 'wurt'). This is done using a process known as *lautering*.

1. Mash out

The first step is to 'mash out'. This involves turning on your heat source to bring the temperature of your grain up to 75°C (167°F). The higher temperature increases the *solubility* of sugar in your wort. Therefore, during the 10 minutes or so that you hold it there, more sugar seeps out of your grain and you get much, much higher efficiency.

If you've got a little pot, mashing out is as simple as placing it on the hob and heating it up. If you've got a big pot with an element, make sure you rig a system where you bring the bag off the element before mashing out. This prevents scorching and underwater fires. I use my stepladder bag suspension system.

Turn your heat source on, and heat until your grain reaches 75°C (167°F). If the wort around your bag is reaching temperatures above 80°C (176°F), turn off the heat and let the temperature settle for a few minutes. Give it a stir, and you can re-assess and turn the heat back on, if appropriate.

2. Recirculate

After you've mashed out, the next step would be traditionally to *recirculate*. This is when you pump your wort from a pipe underneath your mash filter (in our case, the bag) and then over the top of your mash tun and back through your grain. This causes the grain (so long as it's not disturbed) to form a filter of its own and remove a lot of the waste products that fall to the bottom as trub.

Using the BIAB method, recirculating isn't strictly necessary. The truth is, these waste products don't negatively impact your beer in any way, except for their potential to cause scorching at your heat source. If this is a risk you are still worried about, you can recirculate as below:

1. Suspend your grain bag above the wort-level in your pot. Secure the bag using your rope.

2. Take off jugfuls of water using the largest jug you have, then pour them into (or through) your grain, above. You want to recirculate your entire volume of wort at least once.

3. Stop when you notice that your wort is significantly clearer than it was before. Whatever you do, don't upset or stir your grain bag, as this will cause trub to fall into your beer.

3. Sparging

Sparge is maybe my favourite word. This is when you rinse the grains in order to extract every last morsel of sugar, and it is done to increase efficiency. Like recirculating, many people who use BIAB setups miss out this altogether.

Sparging is carried out with water at 75–80°C (167–176°F) – the water should be at least the temperature of your mash-out. Any higher than this and you increase the risk of extracting tannins from the grain husks, which will give your beer an unpleasant astringency.

Lautering

1. Loop your rope around your step ladder and pull taught to bring your bag off the element

2. Tie your bag above the wort-line, so that the wort drains from the grain

3. Remove jugs of wort, pouring through the bag until the wort from the tap is no longer cloudy

4. Sparge by pouring hot water through the bag

Similarly, if you sparge with too much water, the pH of the grain will increase and you'll also extract tannins.

That said, this is very unlikely in BIAB – we only want to sparge with a few litres/quarts, in order to rinse the grain of only the most easily dissolved sugars. After the first few litres/quarts, you need a lot more water and plenty of time to extract anything else that's useable.

If you've got a big setup, lifting out the bag and holding it above your pot is a little more tricky. This is why I recommend a stepladder setup (see the step-by-step photos on the previous page). You want to sparge with about 5 litres/quarts of water, or whatever the recipe says. Blend water from the kettle and cold water at a ratio of 3:1 to make water that's about 75–80°C (167–176°F).

If you've got a mini setup, remove the bag completely from the pot and place it in another pot, food-safe bucket or mixing bowl, before pouring over your hot water.

Remember:

Specific gravity is influenced heavily by temperature. Most hydrometers are calibrated to read at 20°C (68°F). To compensate, measure the temperature of your sample in your hydrometer jar, and enter this into a hydrometer adjustment tool in an app, or on your brewing software.

Step 5: boil

What you actually need to do
- Take a gravity reading
- Bring your wort to the boil
- Boil your wort according to your recipe
- Add hops, near the start for bittering and near the end for aroma
- Add your chiller and finings 15 minutes before the end

Why we boil

A solid 60–90 minutes might seem like a long time for which to boil, but there's good reason why the boil is a beer's basic right. First and foremost, it's there to sanitise your beer. You want to kill the bugs that live within your grain, and boiling is the only way to guarantee this.

It gives a **maltier** beer, causing both caramelisation and Maillard (browning) reactions. It gives a **cleaner** aroma, blowing off nasty aromatics such as dimethyl sulfide (DMS) – that's why you should never put a lid on (even partially) during the boil. It **clears** the beer, as the boiling causes proteins and tannins to clump together in what's called the **hot break**. You'll notice this just at the beginning of the boil – watch, because this is the point at which the beer is most likely to foam over the top. Once the boil begins, it concentrates the beer (so making it **stronger**) by blowing off excess water.

Then, boiling has a few major effects on **hops**. First, it *isomerises* the alpha acids contained in the hops. Isomerised alpha acids are bitter and they are soluble, so they dissolve into your wort and persist through to the finished beer.

Like in the removal of DMS, boiling also removes the aromatic portions of hop flavour compounds over time. Adding hops at the start of the boil is good for adding hop bitterness to beers that aren't supposed to smell strongly of floral or citrus aromas. Adding them near the end is good for hoppy English or American-style beers.

How to boil

1. Turn your heat source on to bring your wort up to the boil over maximum heat. If you've got any first wort hop additions planned, add these as soon as you've turned your heat source on.

2. While your wort is heating up, drain or scoop a little wort off and take a gravity reading using your hydrometer. This gives you your *pre-boil gravity*. You can use this to work out the *mash efficiency* using brewing software.

3. Keep checking the temperature. When it approaches 95°C (201°F) or more, you're nearly at hot break time. If you've not got much space between your wort and the top of your brew pot, you might want to turn down the heat at this point. Or at least watch the pot very carefully – it's about to foam up.

4. At hot break, add your bittering hops (if using) and start your boil timer on your phone or brewing software. Aim for a good, rolling boil to blow off that DMS.

5. Add your hops, at each specified addition. If you're using Irish moss-based finings, such as *Protofloc* or *Whirlfloc*, add these 15 minutes before the end of your boil. If you have a wort chiller, add this at the same time in order to sanitise it.

6. At the end of your boil time, it's important to turn off the heat straight away and get cooling as soon as possible. This is *flameout*, and some recipes will add hops here. Be aware that even though your beer isn't boiling, any hops added will contribute bitterness with every passing minute until it's below 79°C (174°F).

Boiling

1. Measure a pre-boil gravity. Take into account the high temperature

2. Bring to a rolling boil

3. Add hops as your recipe states – use a bag if you don't have a filter

4. Add your chiller and finings 15 minutes before the end

Optional: steep
(for American-style or hop-forward beers)

What you actually need to do:
- Cool your wort down to roughly 75–79°C (167–174°F)
- Add your aroma steep hop addition
- Steep for 30 minutes

If you want your beer to be super-hoppy, you'll need to add more hoppiness, without adding bitterness. Here's how.

Alpha acid isomerisation, causing bitterness, occurs at 79°C (174°F) and higher. The higher the temperature, the faster the reactions happen. At home, we can easily halt the isomerisation of alpha acids by cooling to below 79°C (174°F), before adding our aroma additions. This means we can leave these hops to steep for as long as we like without worrying about adding any more alpha acid bitterness at all. We can even keep the lid on and retain every single last drop of hop aroma.

Conducting this cooling step is easy – just follow the steps for cooling below, and halt the process once you hit about 75–79°C (167–174°F). If you've got an efficient chiller, this could take seconds. Let the wort sit at about this temperature for 20–30 minutes for the hop flavour to infuse.

If you've brewed a batch that you plan on diluting (liquoring back) with cold, near-sterile water from the tap, you can quickly cool to steeping temperature by adding cold tap water at this stage.

Step 6: cool

What you actually need to do:
- Attach the hoses to your chiller and taps
- Turn on and leave to cool to 18°C (64°F)
- Stir to accelerate cooling
- Repeatedly test the temperature

Your wort needs to reach the right temperature for adding the yeast. This is required in order to get a good fermentation. You could just leave your pot to stand, to allow the wort to cool gradually by ambient temperature alone, but there are several reasons why you should not do this.

First, cooling quickly helps prevent infection. Your virgin wort is the ideal culture medium for infective organisms. It's safer from infection once it is cooled and the yeast is added: they grow in huge numbers, use up the sugars and release chemicals that inhibit the growth of other organisms. This is called *competitive inhibition.* It is therefore imperative to minimise the amount of time you leave before you add your yeast.

Cooling your wort quickly causes 'cold break'. Just like 'hot break', these are clumps of protein that form with temperature change. The large size of cold-break clumps causes them to precipitate in a dramatic fashion on rapid cooling (especially in the presence of wort finings like Irish moss), preventing *chill haze*, a cloudiness that only appears when your beer is cold.

Chilling your beer

Before you begin to chill, your beer has been up at the sort of temperatures that kill pretty much any bug on contact. **As soon as you begin to chill, you need to keep everything that touches your beer sanitary**. All equipment, now, needs to be very clean and then sanitised. Have your spray bottle at the ready. I really need to get some sort of holster…

For small batches (<10 litres/quarts)

The simplest way to chill your beer is to place the lid on your pot, then wrap the gap between the lid and pot in plastic wrap. Place your pot in a bath or sink, then run cold water from the tap to create a cold-water bath. If you like, you can add ice. Swirl your pot repeatedly in order to cool down your wort faster, and replace your bathwater and ice as needed.

This technique has the disadvantage that, in order to check the temperature, you have to unwrap the lid and stick a (sanitised) thermometer inside. If your temperature isn't quite down enough, you'll have to rewrap. As a rule, your pot should feel at least tepid to touch before you even think about checking it.

Liquoring back to cool

If you have brewed a really strong batch and you would like to dilute your beer to make a weaker one, you can liquor back (see 'mashing') with cold sanitised water to cool those last final few degrees, which can take a long time as the difference in temperature between your water and your wort is so small.

For larger batches (>10 litres/quarts)

Use a chiller. There are lots of chillers out there, but I recommend a copper immersion chiller.

The easiest way to cool with an immersion chiller is to simply attach the lines to the cold water supply and off you go. However, if you want really fast chilling, you should consider stirring the wort using a clean and sanitised (preferably stainless steel) implement. As the water runs through the copper coils, it rapidly cools the wort that is immediately next to the pipes. But the rest of the wort must still be left to cool by dissemination. If you stir in the opposite direction to the way that the water is flowing through the coil, you can increase the contact between the cold copper and the hot wort.

Cooling when it's hot

If you live in a tropical country with a very high groundwater temperature, you might be unlucky enough to require two copper immersion chillers. These should be connected to one another by an extra bit of hose. One is for dipping in your beer, while the other should be dipped in a bucket full of ice. Your tap water should then run through the ice to cool it down, so that it then chills the beer more effectively. I'm not the best person to teach you about this: I'm from Scotland.

Step 7: pitch

What you actually need to do:
- Use the tap to transfer your beer into your bucket (or pour it)
- Aerate the cool wort
- Add the yeast and combine

'*Pitching*' means to add the yeast. But before you can pitch, you need to make sure your wort is prepared for the yeast. This involves the following steps:

1. Cool down the wort to the right temperature
Before your yeast can be added, your wort must be cooled to an exact temperature depending on your recipe and your yeast strain.

For standard ales, and thus most beers you'll be making, this temperature will be 16–20°C (61–68°F).

2. Check the original gravity
Doing a check on the original gravity (OG, see page 23) is essential at this point. This is your last chance to *liquor back* (dilute) with sanitary water until you hit your numbers. If you're below your intended gravity, I'm afraid there's nothing you can do now – you're going to end up with a slightly weaker beer than you'd planned. It's a lesson to brew slightly stronger next time, because you will always have the opportunity to dilute at this point.

When checking your gravity, do not add the wort from the hydrometer jar back into the main batch, as this is an infection risk. Drink the wort – it's quite nice. It's especially nice with good Scotch whisky.

3. Transfer and aerate the wort
Right before pitching the yeast is the one time in brewing that you want to get as much air into your wort as possible. Yeast need oxygen in order to grow, so we want to oxygenate the wort before adding the yeast. You can do this while the wort is still in your pot, when it's in your bucket or on the way in between.

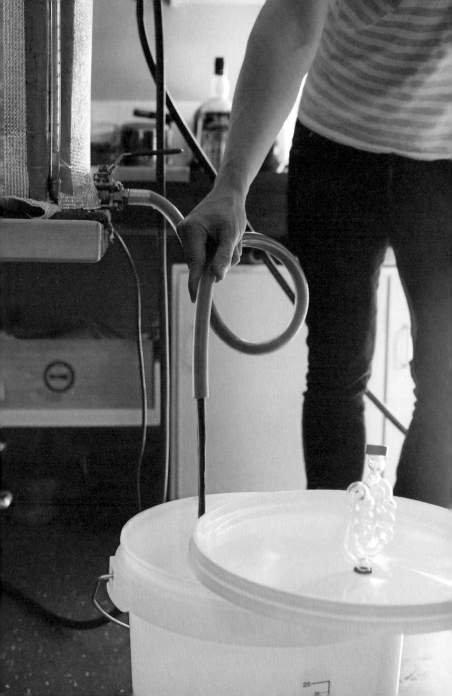

Make sure your bucket, tap, lid and airlock are disassembled, cleaned and sanitised before thinking about transferring.

To transfer a small batch, the easiest way is to pour wort directly from your small pot into your bucket. You've got to do this in one harsh movement, to stop drips down the side of the bucket reaching your beer. As a precaution, you should clean and sanitise the side of your pot too.

For larger volumes, or if you have a tap on your boiler, you should sanitise the tap thoroughly and attach a piece of sanitised silicone tubing to it. You can then direct the other end of the tube into your bucket – it's best to do this from a height for aeration. The more splashing, the better.

Either way, seal the lid on your bucket and begin to rock it backwards and forwards to create as many splashes as you can. If you are strong enough to pick the whole thing up, even better. Just don't drop it, and make sure the lid's on right.

4. Add your prepared yeast

The final step is to add the yeast that you prepared earlier. If you made a starter, I hope you remembered to take it out of the fridge to come to room temperature – a cold–hot temperature change will shock the yeast and result in a stressed-out fermentation.

Just swirl the yeast to make sure it's all in solution, then unclip your lid and pour it in. Once you've replaced your lid, you can give your bucket an extra shake to distribute your yeast and compound your aeration.

You might think that's your work done. But just look at all that mess – get it cleaned **now**. Do not leave it to the morning, and definitely not until the next time you brew. The shorter the time between brewing and cleaning, the easier the cleaning will be.

Step 8: ferment

What you actually need to do
- Keep your beer in a cool, dark place
- Check and control the temperature
- Avoid taking too many gravity readings
- Be patient

Fermentation is the longest and most important stage in beer making. You should expect to leave your beer for about 2 weeks. During this time, your yeast will turn your wort into beer.

Getting a good fermentation

First, you've *pitched enough healthy yeast*, and you've aerated your wort well. Next, through *temperature control*, you have the biggest potential to destroy your beer, or create a masterpiece. If your beer gets wildly too hot, your yeast will first wreak havoc and then it will die. If your beer gets too cold, your yeast will just give up and might never spring back into life.

To get a clean fermentation for most ales, you should keep your fermenter between 18 and 20°C (64–68°F). Handily, this is your average room temperature, and it's why so many more home brewers brew ales and not lagers. You should find the part of your house that stays within this range most consistently, and place your fermenting beer there.

If your yeast get too cold, the problem is simple: under-attenuation. This leaves you with sweet beer. Your beer could then begin to attenuate further in the bottle and cause these to explode.

If your fermentation is too hot, especially during the growth phase in early fermentation, you'll get stressed yeast. They'll make fusel alcohols, causing paint thinner smells. Then, because the beer is hot, the fermentation is faster and produces more heat. This runaway reaction can cause a beer to stall when most of the sugar is used up and the temperature begins to drop.

Cooling down the beer

During the first three days of fermentation (when the yeast is growing), it is especially important to keep your ale's temperature down. Yeast should generally be pitched as close to 18°C (64°F) as possible. If the temperature creeps above 20°C (68°F), start thinking about doing something. If it hits as high as 22°C (72°F), act.

The easiest way to cool your beer is to soak a towel in cold water, then wrap it around your fermenter. Not only does the temperature of the water cool the beer, but its evaporation is an endothermic reaction – it draws heat from its surroundings. This means that as the towel dries, it cools your beer. If you couple your wet towel with an electric fan, you speed up the drying process significantly.

The cooling results from a wet towel, especially combined with a fan, can be staggering. So much so that you want to check your beer doesn't cool down too much and cause your yeast to drop out.

In order to gain a more consistent temperature, it can be a good idea to place your fermenter inside a larger vessel, such as an even bigger bucket or a bath. If you fill this vessel with water at your desired temperature, the heat generated by your beer will diffuse quickly. Being a much larger volume, it will change temperature much more slowly in response to ambient heat or cooling. This keeps it at a consistent temperature, day and night.

Warming up the beer

If you're in the unfortunate position of having a cold, draughty home, you'll need to heat up your beer. You should act if your beer hits 16°C (61°F) or below. The easiest way to do this is to move your fermenter next to a heater, if you've got central heating.

Another way to heat up your beer is to place it in a larger vessel, such as a bath. If you fill this with water that's slightly warmer than your desired fermentation temperature, your beer will creep up gradually.

If you're really struggling to keep the temperature up, you should

consider buying a *fish tank* heater. These are cheap, submersible heaters that you can sanitise and place inside your beer, or preferably in the water-filled vessel in which your beer is sitting. Just sanitise it first.

Conditioning

Once your yeast have finished fermenting the sugars in your beer, their job isn't done. Their next stage is conditioning – this is when they turn to alternative sources of food. Handily, this includes plenty of the nasty by-products of their growth.

Many guides will tell you to transfer your beer to another, *secondary* fermenter for conditioning. Don't. Leave your beer on the yeast. In your primary fermenter, conditioning is more effective (because there is so much more yeast) and you don't have the risks of infection and oxidation from an extra, unnecessary transfer. Two weeks in the primary fermenter is adequate to ferment and condition your beer.

Two common problems caused by cutting short conditioning, or bottling the beer too early, are *acetaldehyde* and *diacetyl*. These yeast by-products make your beer smell of green apples and butter respectively, and both are cleared the longer you leave your beer to condition at appropriate temperatures. Giving your beer extra time also lets yeast and protein settle to the bottom, resulting in a clearer final product.

Knowing when your beer is done

First things first. Never, ever judge the progression of fermentation by the bubbling of an airlock. CO_2 is steadily released from any liquid in which it is dissolved, especially if it heats up. Bubbling does not correspond to yeast activity.

Judge the progress of your fermentation by its current *specific gravity*. **Don't take gravity readings every day** – this is a waste of beer, and it doesn't help you. Only take your first when you think the beer looks like it has finished fermenting. After you have it, check again the next day, and the day after that. **As a rough rule, your beer is done if your gravity remains the same for 3 days in a row**.

Always spray some sanitiser into your tap after removing a sample for a gravity reading. This stops a build-up of impossible-to-clean sticky beer that could harbour infection over time.

Your primary could be complete in anything from 3 days to 2 weeks. When you're starting out, I'd always advocate leaving it on the yeast for a full 2 weeks before bottling. If you've got three identical gravity readings before this time and are anxious to bottle, you should at least leave it an extra few days to complete conditioning.

The only time that testing multiple gravities is not accurate is during a *stuck fermentation*. This is when your yeast, for whatever reason, flocculates even though there are plenty of sugars left. You'll have specific gravities much higher than expected, that won't shift. This can be caused by a sudden drop in temperature, underpitching and/or unhealthy yeast.

Dealing with a stuck fermentation

Yeast are annoying buggers. Once they flocculate, they are difficult to rouse without significantly changing their environment.

The first thing you should try is heating your fermenter – bring it up to the top limit of the yeast's fermentation range. Combine this with a swirl or mix, in order to disturb your yeast bed, and a wee dose (50g/1¾oz, for example) of table sugar. This triple-assault will solve the vast majority of stuck fermentations. Be careful not to splash, or you could end up with not just cloying beer, but oxidised, cloying beer.

If this doesn't help, you're going to have to transfer your beer into a secondary fermenter (see the next page) and re-pitch a standard, healthy quantity of new yeast. You should attempt to let your old yeast settle out as much as possible, in order to leave it behind; yeast can actually produce markers that communicate with your new yeast and tell it to flocculate, too.

Step 9: secondary fermentation & dry hopping (optional)

What you actually need to do:
- Sanitise hop bag
- Weigh dry hops into hop bag, and add these to the secondary fermenter
- Transfer your beer from one bucket into another
- Leave your beer for 3 days at 16–23°C (61–73°F)

If you are brewing a hoppy beer, you might want to include dry hops. These are hops added after fermentation, and contribute a very fresh and powerful aroma.

You can dry hop in your primary fermenter. There are pros and cons to this – hop oils can stick to the walls of yeast, and therefore be dragged down into the trub, meaning you don't get quite the flavour you expected. But the difference is small, so don't worry about it if you don't have a spare bucket.

Transferring to a secondary fermenter removes the majority of the yeast, but it does also introduce inevitable oxygen and an increased risk of infection. Depending on how conscientious you are, it's often safest just to dry hop in primary

You do not need to worry too much about introducing infection from the hops themselves, as hops are antibacterial. They can harbour wild yeasts, but usually in very small quantities. The main risk from infection comes from improperly cleaning and sanitising your secondary equipment, and the use of a hop bag.

A hop bag is exactly what it sounds like – a nylon bag into which you can put hops to contain them. If you're dry hopping with pellet hops, using a hop bag is a good idea. To avoid infection, you should boil the bag in a saucepan before use, as you cannot sanitise such a fine mesh using a sanitiser alone.

If you're dry hopping with whole-leaf hops, just pour them straight into your sanitised secondary fermenter, and *rack* (transfer) the beer on top of them. Be careful not to splash, as this will cause oxidation. Whole leaf hops float, so it's easy to avoid them when you transfer your beer to your bottling bucket.

How to dry hop or transfer

1. Disassemble, clean, sanitise and reassemble your spare bucket, tap and airlock. This is to be your secondary fermenter, in which you will dry hop.

2. If using a hop bag, boil it in a saucepan of water to sanitise it. Weigh your hops into a clean, sanitary container, then place these in your hop bag. Close the bag and place it in the bottom of your secondary fermenter.

3. Place your primary fermentation bucket on a chair, and your secondary bucket on the floor. Your tubing should be able to reach from the tap of the primary to the bottom of the secondary.

4. Sanitise the tap of the primary fermenter, then attach your sanitised silicone tubing.

5. Start the flow slowly, watching not to splash. Tilt the bucket as you start out, in order to keep the end of the tube always submerged. Let it fill to the top, leaving any yeasty sediment behind in the primary.

6. Replace the lid and place your bucket in a dark place for 3 days at room temperature. Because much longer than this can cause a grassy taste to your beer, you should always make sure you have time to bottle your beer after the third day of dry hopping.

7. Clean your primary fermenter and other equipment thoroughly. If you like, you can first wash your yeast and save it for your next brew.

A note on funky beers

If you're even thinking of adding bugs to beer, I probably don't need to convince you of how awesome an idea this is. Sour beers can be so different to what someone might think of when they imagine *beer*. They can be complex, 'funky', tart, sour, barnyardy, fruity, oh-so-dry and all of these at once.

Making sour beers is dead easy, if you follow the basic rules of brewing. I'm assuming you've read most of this book, have a bit of experience in all-grain brewing and are confident in your brewing practice.

The major thing that's different with sour beers is time – the three major bugs we might want in our 'wild' beer are the same three beer pathogens we really don't want to infect our everyday beer – *Brettanomyces*, *Pediococcus* and *Lactobacillus* (see pages 65–66). These tend to grow much slower than brewer's yeast (*Saccharomyces* sp.) and therefore they take time to reveal their character. By time, I'm talking at least a couple months, initially. Depending on the individual circumstance, it could take a year or more for you to hit the flavour you want.

It's a good idea to have a completely separate set of fermentation and transferring equipment for brewing sour beers. When we clean and sanitise our equipment, we do so to remove the potential pathogens that just might have made their way in. We're talking tiny numbers of cells – in the hundreds or the thousands. When we make a sour beer, we're actively encouraging the growth of **hundreds of billions** of these same cells. Even the most conscientious sanitiser, such as myself, cannot be confident of killing all of them.

Which bugs to use

When deciding how to funk up your beers, there are two main methods: *pitching from dregs* and purchasing ready-made *sour cultures*.

Pitching from dregs

Almost all sour or 'wild' beers are bottle conditioned with the same bugs that give them their characteristic flavour. The numbers of each bug change over time, and you will never get anything exactly like the brewery that made the original beer. This doesn't really matter, unless you're trying to make an exact clone. Pitch the dregs from as many bottles of your favourite sour beer as you can, though, and you're still going to get something awesome.

If you want to use dregs (and you really should), you should be sure to ferment alongside a primary strain of *Saccharomyces cerevisiae*. I'd go for a saison yeast, for its distinctive fruity character. At the very least, you should probably go for a Belgian strain, or even a pre-made mix of bugs. Whatever you pick, be aware that almost all of its character will likely be overwhelmed by the bugs from your bottles. When your primary yeast starts to settle, they will prosper.

Pitching from pre-made cultures

The explosion in popularity of sour beers means that there are now many excellent blends of bugs, ready to pitch without the need for yeast starters. Saison-Brett blends are my favourites, especially supplemented with some dregs, or my house culture. I recommend looking at The Yeast Bay, WHC, Omega Yeast Labs, Bootleg Biology and Mainiacal Yeast.

Milk the Funk

A special mention has to go to the Milk the Funk community, who have put together an amazing Wiki with such in-depth information about making sour and funky beers. If you're interested, check it out: milkthefunk.com.

10-litre brew day

1	4	7	10	13
2	5	8	11	14
3	6	9	12	15

1. PREPARE YOUR YEAST
2. MASHING IN
3. STOVE-TOP MASH-OUT
4. MINIATURE LAUTERING
5. FIRST RUNNINGS
6. SPARGING WITH A KETTLE
7. LEAVE GRAINS TO SOAK
8. SQUEEZE THE BAG
9. TAKE A PRE-BOIL
   GRAVITY TEST
10. ADD FWH AND START BOIL
11. ADD FLAVOUR HOPS
12. KETTLE FININGS
13. AROMA STEEP
14. LIQUOR BACK WITH
    SANITARY WATER
15. COOL IN A COLD BATH

Bottling & storing beer

Preparing your bottles

This is the bit they kept quiet about. If you're a precisionist who likes methodically going through really boring things for the sake of nothing, you're going to love bottling.

The rest of us? After the first few times, we realise how rubbish it really is. It's the perfect anti-climax – loads of hard work with little or no reward for still weeks to come. I'd recommend getting a friend to come along and help – it makes it much more bearable to share the burden. Besides, it's an excuse to open a few beers.

Cleaning (and de-labelling)

The first step is cleaning your bottles. Make sure you've got enough – you'll need roughly 40 × 500ml (1 US pint) bottles, or 60 × 330ml (11fl oz) bottles, for a 20-litre/quart batch. More likely, you'll have collected a combination of the two. Most pubs, if approached, will happily point you in the direction of their recycling bins if you're running short.

If you've got a dishwasher, you're golden. Put your bottles in the dishwasher on the hottest wash. Your bottles are now clean and sanitised (though I would still spray with sanitiser – see page 116).

Otherwise, you'll need to soak your bottles in warm soapy water, ensure they are absolutely clean on the inside. At the same time, you can scrub and scrape the labels off – a butter knife helps. Once de-labelled, rinse each bottle with water until there is no foam residue left inside.

Sanitising

This is pretty easy – hold your bottle in one hand and your spray bottle in the other. Hold the bottle at an angle, and spray six or seven times into it, turning it as you do so. You want to make sure that you have covered all the inner surfaces in your sanitising solution. Set out your sanitised bottles in rows, ready to be filled. Leave the sanitiser in there, for now.

Count out the number of caps you need and place these in a bowl. Try not to have too many left over, as the contact with the sanitiser will likely make them rust before you bottle your next batch. Cover these with sanitiser. If you have swing-tops, you'll want to spray the rubber rings with plenty of sanitiser, too.

Priming

Before you bottle your beer, it's worth checking how strong it is. The formula only requires the original gravity (OG) and the final gravity (FG):

$$\text{Alcohol by volume \% (\%ABV)} = (OG - FG) \times 131.25$$

This will give an approximation of your alcohol content. I almost never do it by hand as I never remember it. And I'm lazy – I've got about six different apps that do it more accurately. You should get one, too.

The next stage is priming. This refers to the process of adding sugar to your beer before bottling. The remaining yeast will ferment this sugar, creating CO_2 (as well as a wee bit more alcohol). Because this CO_2 cannot escape, the pressure inside the bottle builds up, and you end up with fizzy beer.

Start by cleaning and sanitising your bottling bucket to the same meticulous extent as you did your fermentation bucket. You'll also want to find and clean your *silicone tubing*, your *bottling stick* and a *measuring jug*.

Place your measuring jug on a set of scales, and weigh your required sugar (see carbonation chart). Into your jug, pour at least 300ml/11fl oz of just-boiled water, and stir with a sanitised spoon to dissolve the sugar. This is your *priming solution*. Pour away any sanitiser left in your bucket, and then pour your priming solution in instead.

Place your bottling bucket, with the lid placed loosely over it, on the floor. Lift your full fermenter onto a table or chair, so there's a height difference between them. You'll probably want to leave this still for 5 minutes for the yeast to settle. Meanwhile, sanitise your silicone tubing by spraying inside it and ensuring foam touches the entire inner surface, before sanitising the outside too. Sanitise the fermenter's tap. In fact, just spray everything with your sanitiser.

Attach your tubing to your tap, and place the free end of the tube into the priming solution so that it touches the bottom of your bottling bucket. Keep your bottling bucket mostly covered with its lid, so as to stop any bacteria-laden dust falling into it.

Open your tap, and let your beer flow gently down from one bucket to the other, keeping the end of the tube submerged. The flow from the tube should adequately mix your beer with the priming solution. Whatever you do, though, *don't splash*, as this introduces oxygen into your beer. You might want to tip your fermenter slightly as the beer runs dry, but don't so much as to disturb the layer of yeast and sediment (we call this *trub)* at the bottom. The trub should all be left behind. Once you've got all your beer out, remove the tube from the bottling bucket and half clip its lid shut. Move your fermenter out of the way and place your bottling bucket onto a work surface. It's finally time to bottle your beautiful beer.

Priming

1. Measure your sugar, then make a solution with boiled water

2. Sanitise your silicone tubing with plenty of sanitiser

3. Place your fermenter on a table and your bottling bucket on the floor

4. Syphon your beer from your fermenter into your bottling bucket, using a tap as necessary

Bottling

This is where you start your very own production line. In front of you, on a table or a work surface, you should have your full bottling bucket, lid covering the beer but not sealed. Next to it, you should have enough sanitised bottles to do the whole batch. On the other side, I have my bowlful of crown caps and my crown capper.

The tap of your bottling bucket should poke over the edge of your work surface. To this, attach your (cleaned and sanitised) bottling stick. (Remember, you might need a bit of tubing, and possibly even clip, for your bottling stick to be compatible with the tap.) Open the tap, and watch the beer fill the stick.

Place a bowl directly underneath the stick – this will catch most of the drips and be somewhere to pour any excess sanitiser.

Take your first bottle, and pour its remaining sanitiser into the bowl – this will re-sanitise your bottle and its neck. Place the bottle underneath the bottling stick so that it begins to fill – watch it carefully. When the beer reaches the very top of the bottle, remove it. Once removed from underneath the bottling stick, you want your bottles filled to at least the start of their necks. Place a sanitised cap on top, and seal it using your capper.

Congratulations. Admire that bottle. It should feel good. Just a million more to go.

As you have less and less beer in your bucket, you're going to want to tip the bucket so the tap is always filling with beer – I like to use a rolled towel to prop the back side of my bucket up. Be careful not to disrupt any sediment, lest you have one final bottle of yeasty trub. If you don't keep the tap covered, you'll draw oxygen down into your bottles, causing oxidation.

Bottle conditioning

Once your beer is bottled, you'll want to mark the bottles in some way. I use a permanent marker to write an identifying word or two and the alcohol by volume on the cap. Some people identify their beers using only different coloured caps, and some people go as far as designing and printing labels for each beer. It's totally up to you.

Move your bottles to a place away from direct sunlight and at roughly room temperature. This stage is called *bottle conditioning*, and it does just that: it conditions. Not only will your beer carbonate in the bottle, but your yeast will turn to food sources it would otherwise turn its nose up at. These include inevitable off flavours. They will also mop up most of the oxygen you introduced when bottling, preventing oxidation. **Condition your bottles for 2 weeks before judging your finished beer**. You might well have a drinkable, carbonated beer in 3 days, though, if you're lucky.

Don't store the bottles in the fridge or anywhere cold – this will arrest the secondary fermentation that happens in the bottle and leave you with flat beer. Thankfully, this second fermentation isn't quite so critical. When yeast metabolises pure sugar, it doesn't produce quite so many off flavours at warmer temperatures. Thus if you want your beer to carbonate quickly, keep the bottles in a warm place.

Make sure you clean up, and properly. There is nothing worse than sticky beer-mess encrusted on your equipment the next time you want to brew. As a rule, you should never put away any piece of equipment dirty.

Carbonation

Different beers work better with different levels of fizziness. We can control how fizzy our beers are by changing the amount of sugar we add before bottling. We measure this by the amount of carbon dioxide (CO_2) that's dissolved in the beer – the **volumes** of CO_2.

I wish it was as easy as 'this much sugar gives this much fizziness'. As well as the volume of the beer, the temperature impacts how much sugar you should add. If your beer is colder, more CO_2 will be dissolved in it and thus you don't need to add so much sugar. If it's warmer, the opposite is true, as CO_2 comes *out of solution* as liquids warm up. This is exactly why a warm carbonated bottle might gush with foam, while a cold one lets off nothing more than a gentle hiss.

Below are the carbonation levels for various beer styles: 1.5 volumes is very mild, 2.5 volumes is spritely, and anything over 3 volumes is likely to foam if opened at room temperature, or even lightly shaken up. Most bottles will take up to 4 or 5 volumes, but watch out for those bottles that seem thin or fragile; they most likely are.

Carbonation levels

British pale ales and bitters	1.6–2 volumes
Porters and stouts	1.8–2.2 volumes
Belgian ales (pale)	2.5–3.5 volumes
Belgian ales (dark)	2–2.5 volumes
Saisons and farmhouse ales	3.5+ volumes
American ales	2.2–2.7 volumes
Lambics	2.5–3.5 volumes
Lagers	2.2–2.7 volumes
Wheat beers	3.5–4.5 volumes

For best results, you should use an online or app *priming sugar calculator,* but below is a chart showing roughly how much table sugar (caster sugar; granulated sugar) you should add for each 20 litres/quarts of beer, to reach certain volumes. You can halve the amounts for 10-litre/quart batches.

Priming sugar calculator – 20-litre/quart batch (about 18 litres/quarts in bottling bucket)

	1.6	1.8	2	2.2	2.4	2.6	2.8	3	3.2
16°C	50g	66g	82g	98g	114g	130g	146g	162g	178g
17°C	52g	68g	84g	100g	116g	132g	148g	164g	180g
18°C	55g	71g	87g	103g	119g	135g	151g	167g	183g
19°C	57g	73g	89g	105g	121g	137g	153g	169g	185g
20°C	59g	75g	91g	107g	123g	139g	155g	171g	187g
21°C	61g	77g	93g	109g	125g	141g	157g	173g	189g
22°C	63g	79g	95g	111g	127g	143g	159g	175g	191g

Alternatives to bottling

Yes, bottling takes ages. It can be a pain in the arse. You'll probably end up with a sticky floor and a bath plughole full of shredded beer labels. However, I'd still recommend it. Wholeheartedly. Like brewing, it takes a lot of time but it is rewarding – there really is nothing like the whoosh when you bend back the cap on that first carbonated bottle.

Plastic kegs

The first option you're likely to encounter is the pressurised plastic keg system, the ones with little CO_2 canisters. Just don't. You'll have leaks aplenty, flat beer, and burn through your canisters. These plastic kegs are for those who don't know about beer, don't really care about their beer. Avoid them.

Minikegs

Minikegs are 5-litre/quart steel kegs. They have a hole in the top covered by a rubber bung and they often have a plastic tap at the bottom. You might find them filled with branded beers in supermarkets, in which case you can drink the beer, save them and refill them with your own. They are perpetually reusable, unless scratched or deformed.

You might have heard them called 'party kegs'. You can easily and quickly prime and fill them from the top, just like one big bottle. Indeed, it's not a bad idea to fill one or two at the same time as you bottle the rest of the batch. You can dispense with no additional equipment, using the tap at the bottom. They're small enough to fit in the fridge. They're ideal for a night or a weekend between quite a few friends. But as soon as you open the top, oxygen is introduced, so you'll have to drink that 5 litres within a few days.

Full-size keg systems

Kegging is something I resisted for so, so long. I didn't brew beer to sit and drink myself, I thought. I brewed beer to share and speak over. But then I tried it. I saw how awesome having draught beer on tap in your own house was. I'm not including a full draught system setup in this book – if you want one, there are myriad ways of going about it. Any reputable home brew retailer should be able to sell you what you need. Expect to pay about £200 all in, plus the cost of a fridge ('kegerator') or beer line chiller. You'll also need to find a local supplier of CO_2 cylinders and refills; the best way to find this is to join a local home brew group.

Cans

Home canning is a thing. It will cost you a fortune and is extremely cool, but basically pointless. Only consider it if you've got money to burn.

Trouble-shooting your beer

Each beer flaw can be categorised by a technical term, but I'm going to order them by what they smell, taste or look like. The most important thing is to learn to recognise them. Only once you know they are there can you change something in your brewing process in order to improve it. Think about them, one by one, every time you taste one of your beers for the first time.

Buttery; margarine; butterscotch cakey; greasy mouthfeel

Problem: *Diacetyl*
Causes: 1. Inadequate conditioning at appropriate temperature.
2. Infection (*Pediococcus* is particularly notorious).
Prevent: 1. Use plenty of healthy yeast, and allow fermentation temperature to increase after the first 3 days or so. 2. Better cleaning and sanitation.
Fix: Place all the bottles in a warm place (20–25°C/68–77°F) and keep them there – the yeast will metabolise every bit of it they can.

Green apples; cidery smell

Problem: *Acetaldehyde*
Causes: 1. Inadequate conditioning (removing from yeast too early).
2. Infection.
Prevent: 1. Full two weeks conditioning inside the primary fermenter. Use healthy yeast, and well-aerated wort. 2. Better cleaning and sanitation.
Fix: In the fermenter, try adding more yeast and agitating the beer. Some *Brettanomyces* can metabolise, if your beer will work funky. In bottle, it is unlikely to condition out (permanent).

Corn; vegetal; cooked cabbage; celery

Problem: *Dimethyl sulfide (DMS)*
Causes: 1. Inadequate boil (either length or vigour). 2. Using lager malt.
3. Very slow cooling 4. Infection.
Prevent: 1 & 2. A rolling boil for at least one hour, or 90 minutes for extra-pale or lager malts. 3. Speed up wort chilling. 4. Cleaning and sanitation.
Fix: Permanent.

Paint thinner; alcoholic; hot; harsh; acetone

Problem: *Fusel alcohols*
Causes: 1. Higher alcohols produced during rapid yeast growth at higher temperatures. 2. Infection.
Prevent: Keep your fermentation temperature down, especially during the first 3 days. Make sure your wort has been well aerated.
Fix: During extended periods of conditioning, some fusels can be converted to esters, giving a fruity flavour. A mixed-fermentation beer will help.

Meaty; oniony; burnt rubber

Problem: *Yeast autolysis (yeast death)*
Causes: 1. Too long in primary fermenter (months). 2. Huge changes in temperature, especially warmth.
Prevent: Remove your beer from the yeast bed before these flavours develop.
Fix: Permanent.

Sour; tart

Problem: *Infection (Pediococcus or Lactobacillus infection, though Brettanomyces can also cause some sourness)*
Cause: Poor sanitisation.
Prevent: Use a stringent cleaning and sanitisation routine.
Fix: Permanent. But ensure your beer is indeed sour, and becoming more so, before dumping it; some people confuse bitterness and astringency with sourness.

Vinegar

Problem: *Acetobacter infection*
Cause: Poor sanitisation AND oxygen combined.
Prevent: Re-look at fundamental brewing practice: stringent cleaning and sanitisation routine, with better transferring practices (no bubbles); replace plastic brewing equipment; deep clean the rest.
Fix: Wait and see if beer turns into useable vinegar.

Astringent; puckering; grape-skins; over-brewed tea

Problem: *Astringency (note: easily confused with both bitterness and sourness)*
Causes: 1. Over-sparging, or at too high a temperature, extracting tannins. 2. Additional ingredients.
Prevent: Lauter appropriately. Taste the run-off from your grain and ensure it isn't astringent before adding more sparge water.
Fix: Ages out over years. Consider adding *Brettanomyces* or mixed culture and aging.

Sweet; cloying; syrupy

Problem: *Under-attenuation*
Causes: 1. Numerous yeast-related factors including high flocculation, underpitching, sudden drop in temperature during fermentation. 2. High mash temperature.
Prevent: 1. Pitch lots and lots of healthy yeast. 2. Ensure correct mash temperature.
Fix: In bottle, watch for over-carbonation. In fermenter, transfer to new fermenter, leaving old yeast behind, and pitch new, active yeast. Consider different variety.

Skunk; gas-leak; bad breath

Problem: *Skunked*
Cause: Exposure of beer to ultraviolet light.
Prevent: Ferment and store bottled beer in a dark place. Bottle in dark brown bottles.
Fix: Permanent.

Sulphur

If you'd say your beer smells consistently sulphurous, rather than like skunk, you might think of infection (always) as a possible cause. Most sulphur-containing compounds are very volatile, though, and are produced by many lager yeasts. They normally just blow away through the airlock over time.

Cardboard; sherry; musty; soy sauce

Problem: *Oxidised*
Cause: Oxygen introduction during transferring, aging or bottling practices.
Prevent: Minimise transfers between containers, and avoid splashing. When bottling, fill from the bottom, ensure priming sugar is added and minimise headspace.
Fix: It just gets worse.

Hay; grassy; fiery

Problem: *Hop debris or hop overexposure*
Causes: Dry hopping for too long and at too low a temperature.
Prevent: Dry hop for only 3 days at normal fermentation temperatures. Avoid transferring hop debris.
Fix: It will settle over weeks, though not completely. Chill bottles before pouring, leaving the sediment behind.

Gushers; exploding bottles; too fizzy

Problem: *Over-carbonated*
Causes: 1. Intentional in some styles (e.g. Saison). 2. Under-attenuation (see previous page). 3. Too much priming sugar.
Prevent: 1. Make sure there are three days of identical gravity readings, before bottling. 2. Disperse priming sugar through beer (not too syrupy a liquid). 3. Use a priming sugar calculator that takes into account the temperature of your beer.
Fix: Chilling the bottles combined with handling them carefully prevents foaming initially. If severe, you can uncap to relieve some pressure and let them rest for an hour or so for the dissolved CO_2 to release into the air. After this, you can re-cap them and put them back into storage.

Bottle bombs

The other significant cause of gushing bottles is *Brettanomyces* infection (see page 65). Because this wild yeast is slow growing, it can take a while for you to notice it. It will gradually metabolise all the stuff that your yeast cannot, and even move on to munching through dead yeast. This causes huge pressure and proper bottle bombs. Make sure you sanitise everything, including your bottles, expertly.

Spicy; clovey; smoky; plasticky; medicinal

Problem: *Phenolic*
Causes: 1. Intentional character of Belgian and all Hefeweizen yeasts. 2. Severe yeast stress. 3. *Brettanomyces* infection. 4. High water-chlorine content (specific TCP smell)
Prevent: 1. Choose the right yeast. 2. Control fermentation temperature and pitch appropriate numbers. 3. Good sanitisation practices. 4. Treat water with 1 x Campden Tablet per 20 litres/quarts
Fix: May age out over time, especially in mixed fermentation beers.

Thin; watery; light bodied

Problem: *Likely over-attenuated*
Causes: 1. Low mash temperature in a low-ABV beer 2. High fermentation temperature 3. Using too high a sugar content. 4. Infection.
Prevent: 1. Mash at a higher temperature for a fuller body. 2. Cool your fermentation. 3. Reduce sugar content in beer 4. Good sanitation practices.
Fix: Good carbonation can fix thinness to a degree.

No head; looks flat but isn't

Problem: *Poor head retention*
Causes: 1. Dirty glassware. 2. Fat-containing ingredients. 3. High-alcohol beers. 4. Infection.
Prevent: 1. Clean all glassware. 2. Don't add adjuncts containing fat.
3. Use oats or wheat to add body and strength to your head. 4. Improve sanitisation practices.
Fix: Permanent. Or just clean your glass.

Hazy; cloudy; opaque

Problem: *Poor clarity*
Causes: 1. Chill haze (if only present when cold), 2. Acceptable causes such as high hopping levels or a mildly flocculent yeast strain. 3. No finings.
4. High wheat, oat or rye content. 5. Infection.
Prevent: 1. Chill your wort quickly after the boil or steep. 2. Reduce dry hop; use a more flocculent yeast strain. 3. Add finings into the boil.
4. Reduce unmalted adjuncts. 5. Good sanitisation.
Fix: In bottle, chill and keep still. In fermenter, add a leaf of gelatine dissolved in just-boiled water.

Burnt; unpleasant smokiness; acrid

Problem: *Scorching*
Cause: Debris or sugar on your heating element or at heat source; usually in high-ABV beers.
Prevent: Recirculate through your grain bag until clear, or use a coarser crush on your grain.
Fix: Permanent.

1. Scorched

2. Cloudy

3. Headless

4. Syrupy, cloying

Recipes: the standards of brewing

I've laid out some recipes to get you going. Most of these have won some national award or are based directly upon a world-class beer we've both probably tried. They are standards: reliable and brilliant and I'd be happy to have 60 bottles of any one of them in my cupboard.

You'll find more recipes in abundance online: my favourite resource starting out was the database of the podcast Can You Brew It? – just search for 'can you brew it database' and you'll find a good post on homebrewtalk.com. This list will let you clone some of the most renowned commercial beers out there. Be wary of community recipes, posted up on Brewer's Friend or the Grainfather Community, for example: many are just wild. For sour or funky beer recipes, there's no better resource than Milk the Funk.

English pale ale (page 146)

On paper, a more simple beer does not exist than this one. This type of recipe makes a SMASH beer – Single Malt And Single Hop. SMASH beers are designed to showcase quality ingredients, and if you want to experience good British malt, hops and yeast, look no further. If you want a traditional English-style 'Bitter', you can make this recipe but add 400g/14oz of British Crystal Malt to the mash. Keep everything else the same, and you'll get a variant that's ruby-red and slightly sweet with a hint of caramel.

Dry Irish porter (page 148)

A porter is somewhere in between a stout and a brown ale. This is one hell of a delicious beer: my recipe is loosely based on the famous 'Taddy Porter' recipe from Samuel Smith's in Tadcaster, but with a more mellow roasted flavour and the use of an Irish ale yeast. This is a totally underrated yeast that goes so far beyond the Irish stout – it gives brilliant complexity and is a good attenuator. It would be a great choice for a 'house yeast'.

Imperial stout (page 150)

I've blind tasted this beer against some of the great imperial stouts of the world, and it's consistently come out on top. This isn't sweet and cloying like the modern monsters of the imperial stout game: it is distinct in that it is dry. That makes it dangerously drinkable, even at well over 9% abv. Dry does definitely not mean thin – the booze keeps it coating the mouth as you get layer after layer of malt complexity shining through. Pitch plenty of yeast.

Strong Scottish ale (page 152)

This traditional, moderately spiced Scotch ale is a beer you want to brew. It is dark and it is sweet and it is sumptuous. In the recipe you'll notice crushed coriander seeds: lots of hops really don't go that well in Scotch ales, so adding spice gives it a bit of extra complexity in the flavour and a zesty finish. The yeast you use should be of primary concern – Edinburgh or Scottish Ale Yeast is highly advised. These yeasts have a slightly phenolic, Belgiany character to them if you let the temperature creep up during fermentation. Ferment it hot, and let it age for cold evenings of future winters.

Dubbel (page 154)

Not overly strong, but by no means feeble, the distinctive taste of these dark beers comes from two parts: the dark 'candi' sugar and the classic abbey-style yeast. They combine to give a spicy beer filled with dark fruit – plums and raisins more than any.

Though a traditional Belgian 'abbey' style, there's nothing holy involved in making good beer. Meaning you can make just as good a Dubbel at home using nothing more than those crafty monks do. Stick to the simple recipe, and take care of your fermentation by keeping it low for the first couple of days, then letting it rip. Use a heater to maintain it at its peak.

Saison (farmhouse ale) (page 156)

Saison yeasts are some of the most characterful and distinct – you'll have flavours that are earthy, spicy and exceedingly fruity. Saison means 'season' in French, so-called because it was brewed during the off-season for the French-speaking farmers of southern Belgium, then stored for quenching the farmers' thirst during high season.

For a saison to work it must be bone dry. For the most reliable saison, use Danstar Belle Saison yeast, as this strain is extremely attenuative. Some yeasts can be temperamental, stalling out halfway through fermentation. It's for this reason that you first want to let the fermentation temperature creep up and up and up, in order to get good attenuation. You want this beer seriously hot – as high as 32°C (90°F). Then, for true complexity, I add a *Brettanomyces* blend and age for a month in the fermenter, and as long as possible in the bottle.

East Coast IPA (page 158)

This 'juicy' IPA focuses on the floral and fruity hop flavours. And boy, we're using a lot of hops. No finings are added, keeping all possible flavour-carrying compounds in suspension and making sure the mouthfeel is up to scratch. It's imperative you get good attenuation, with a yeast that will take your beer pretty dry. Some of the flavourful English-style yeasts used in this style can be a bit temperamental, so don't be afraid to add a second attenuative yeast, such as WLP001 or US-05.

I've recently taken to using a Kveik yeast for brewing this style – Hornindal or Voss will both work well. Pitch at 35°C (95°F – yes, that hot) and hold it there, and you'll have an amazing, dry, flavourful hop-forward beer in just a few days.

Session IPA (page 160)

Sometimes, a pint is what is called for. Using a complex grain bill, a high mash temperature and an English yeast, we can make a weaker beer that retains plenty of body and avoids becoming thin, as plagues many session IPAs. Despite its diminutive strength, this beer can carry the flavour of a weight of hops meant for beer double its strength.

Sour ale (page 162)

This mixed-fermentation masterpiece is an opportunity to craft your own 'house' culture, for this is how I started mine. Its primary yeast was a saison-Brett blend. But this accompanied a cacophony of dregs from several other beers, including yeasts from Dupont, De Dolle, Fantome and Orval. I've kept this culture going, much like a sourdough starter. It's evolved a wee bit since – I think there's some Cantillon and Hill Farmstead in there now.

The best thing about this is its drinkability. At a mere 4.4% abv, it was the perfect summer beer. I kegged my first attempt, and it disappeared in a matter of weeks.

California common ale (page 164)

This is a hybrid, somewhere between an ale and a lager. If you like lagers, you're going to want to make this beer. It's like an amber lager, but with an ale yeast. It's darker than most lagers, with a very distinctive taste like freshly baked bread. My choice of using British Maris Otter instead of the traditional American two-row adds even more malty character.

The suggested yeast, isolated from Anchor Brewery in San Francisco, is a wonder. Alternatives that give a lager-like character include Kolsch yeasts, or if you want something fermented quickly and you can maintain fermentation temperatures of about 35°C (95°F), the Oslo Kveik strain is very good.

Tripel (page 166)

While a dubbel and a quadrupel are both dark beers full of dark fruit, you might expect a tripel to be of a similar ilk. Nope, this one's golden, spicy and with a little more bitterness and hop presence.

A good tripel is sensational. The highly fermentable gravity yields a boozy beer that coats the mouth and puts the yeast under more stress than most. This allows the Belgian Abbey yeast strains to thrive, as they produce an amazingly complex ester and fusel alcohol profile. My favourite is Westmalle yeast, WLP530 or Wyeast 3787.

American pale ale (page 168)

Many think of the American pale ale as a pale, hop-focused beer: 'just a weaker IPA'. To do so would be unwise – an American IPA is designed to deliver an absolute hop bomb at the expense of everything else. An APA isn't necessarily any weaker than your average IPA. It must be drinkable, yes, but it must also have body and a good amount of malt character. My recipe takes tips from the most celebrated examples in the United States – I've kept it basic with just one hop, Citra. It's awesome, but Centennial would work too.

English pale ale

Target Numbers:

Original gravity	1.042–1.044
Final gravity	1.007–1.011
ABV	4–4.4%
Bitterness	32 IBUs
Colour	7 EBC

Batch size (in fermenter)	20 l/qt
Estimated efficiency	70%

Grain Bill

Premium English Pale Malt, such as Maris Otter or Golden Promise	100% – 4kg/9lb

Hops

Challenger (7.5% AA)	First wort hop – 20g/¾oz
Challenger (7.5% AA)	Boil 15 mins – 40g/1½oz
Challenger (7.5% AA)	Add at flameout – 40g/1½oz

Yeast

Yorkshire Ale yeast, such as Wyeast 1469 or WLP037
Alternatives: English Ale Yeast, such as White labs WLP002, Wyeast 1968 or Safale S-04

Additional Ingredients

1 Protofloc (Irish Moss) tablet

Method

Prepare your yeast. Clean and prepare your brewing equipment.

Bring 20 litres/quarts of water up to 69°C (156°F).

Mash in. Maintain a mash temperature of 65°C (149°F) for 60 minutes.

Mash out – raise your grain temperature to 75°C (167°F).

Sparge with 4 litres/quarts of water at 75°C (167°F) to reach your pre-boil volume of no more than 22 litres/quarts.

Add your first wort hops. Boil your wort for 60 minutes. Add your hop additions at 15 minutes before the end of the boil and at flameout.

Chill your wort to 18°C (64°F). Measure your original gravity. Liquor back with sanitary water to reach your intended OG.

Transfer your wort to a clean and sanitary fermenter. Aerate your wort and pitch your prepared yeast.

Ferment in primary fermenter at 18–20°C (64–68°F) for 2 weeks.

Bottle with 90g/3¼oz of white table sugar to reach 1.9–2.1 volumes of CO_2.

Dry Irish porter

Target Numbers:

Original gravity	1.062–1.066
Final gravity	1.010–1.014
ABV	6.4–6.8%
Bitterness	50–55 IBUs
Colour	20 EBC

Batch size (in fermenter)	20 l/qt
Estimated efficiency	65%

Grain Bill

Pale Malt, Maris Otter	77.8% – 3.5kg/7¾lb
Crystal Malt (80L)	8.9% – 400g/14oz
Chocolate Malt	6.7% – 300g/10½oz
Brown Malt	4.4% – 200g/7oz
Black Treacle (added during boil)	2.2% – 100g/3½oz

Hops

East Kent Goldings (5% AA)	First wort hop – 30g/1oz
East Kent Goldings (5% AA)	Boil 15 mins – 30g/1oz
East Kent Goldings (5% AA)	Boil 1 min – 20g/¾oz

Yeast

Irish Ale Yeast, such as WLP004 or Wyeast 1084
Alternatives: Dry English Ale Yeast, such as Mangrove Jacks m07, WLP007 or Wyeast 1098

Additional Ingredients

1 Protofloc (Irish Moss) tablet

Method

Prepare your yeast. Clean and prepare your brewing equipment.

Bring 24 litres/quarts of water up to 71°C (160°F).

Mash in. Maintain a mash temperature of 66.5°C (152°F) for 60 minutes.

Mash out – raise your grain temperature to 75°C (167°F).

Sparge with 4 litres/quarts of water at 75°C (167°F) to reach your pre-boil volume of no more than 23 litres/quarts.

Add your first wort hops. Boil your wort for 60 minutes, adding your treacle at the beginning. Add your hop additions at 15 minutes and 1 minute before the end of your boil.

Chill your wort to 18°C (64°F). Measure your original gravity. Liquor back with sanitary water to reach your intended OG.

Transfer your wort to a clean and sanitary fermenter. Aerate your wort and pitch your prepared yeast.

Ferment in primary fermenter at 18–20°C (64–68°F) for 2 weeks, or until you have three identical gravity readings over 3 days.

Bottle with 90g/3¼oz of white table sugar to reach 2.0–2.2 volumes of CO_2.

Imperial stout

Target Numbers:

Original gravity	1.085–1.089
Final gravity	1.016–1.020
ABV	9.2–9.5%
Bitterness	85 IBUs
Colour	85 EBC

Batch size (in fermenter)	20 l/qt
Estimated efficiency	65%

Grain Bill

Pale Malt, Maris Otter	80% – 7kg/15½lb
Chocolate Malt	8% – 700g/1½lb
Crystal Malt	4% – 350g/12¼oz
Brown Malt	4% – 350g/12¼oz
Amber Malt	4% – 350g/12¼oz

Hops

Columbus (CTZ) (14% AA)	First wort hop – 50g/1¾oz
Columbus (CTZ) (14% AA)	Boil 10 mins – 30g/1oz

Yeast

West Coast American Ale Yeast, such as US-05, WLP001 or Wyeast 1056

Additional Ingredients

1 Protofloc (Irish Moss) tablet

Method

Prepare your yeast – make sure you have plenty of yeast. Clean and prepare your brewing equipment.

Bring 26 litres/quarts of water up to 72.5°C (162°F).

Mash in. Maintain a mash temperature of 66°C (151°F) for 60 minutes.

Mash out – raise your grain temperature to 75°C (167°F).

Sparge with around 8 litres/quarts of water at 75°C (167°F) to reach your pre-boil volume of no more than 24 litres/quarts.

Add your first wort hops. Bring your wort to a boil then boil for 60 minutes. Add your hop addition at 10 minutes before the end of your boil.

Chill your wort to 18°C (64°F). Measure your original gravity and liquor back with sanitary water to reach your intended OG.

Transfer your wort to a clean and sanitary fermenter. Aerate your wort and pitch your prepared yeast.

Ferment in primary fermenter at 18–20°C (64–68°F) for 2 weeks, or until you have three identical gravity readings over 3 days.

Bottle with 120g/4¼oz of white table sugar to reach 2.4–2.6 volumes of CO_2. Age in the bottle for at least 2 weeks at room temperature.

Strong Scottish ale

Target Numbers:

Original gravity	1.082–1.084
Final gravity	1.019–1.023
ABV 8–8.4%	
Bitterness	26 IBUs
Colour 36 EBC	

Batch size (in fermenter)	20 l/qt
Estimated efficiency	65%

Grain Bill

Pale Malt, Maris Otter	91.5% – 7.5kg/16½lb
Dark Crystal Malt (120L)	7.3% – 600g/1⅜lb
Roasted Barley	1.2% – 100g/3½oz

Hops

East Kent Goldings (5% AA)	First wort hop – 50g/1¾oz

Yeast

Edinburgh or Scottish Ale Yeast; WLP028 or Wyeast 1728
At a push: Dry English or American Ale Yeast, such as Safale US-05

Additional Ingredients

25g/⅞oz coriander seeds, crushed
1 Protofloc (Irish Moss) tablet

Method

Prepare your yeast. Clean and prepare your brewing equipment.

Bring 26 litres/quarts of water up to 72.5°C (162°F).

Mash in. Maintain a mash temperature of 66.5°C (152°F) for 60 minutes.

Mash out – raise your grain temperature to 75°C (167°F).

Sparge with around 6 litres/quarts of water at 75°C (167°F) to reach your pre-boil volume of no more than 24 litres.

Add your first wort hops, then bring your wort to a boil and boil for 60 minutes. Add your crushed coriander addition at 5 minutes before the end of your boil.

Chill your wort to 18°C (64°F). Measure your original gravity and liquor back with sanitary water to reach your intended OG.

Transfer your wort to a clean and sanitary fermenter. Aerate your wort and pitch your prepared yeast.

Ferment in primary fermenter for 18–20°C (64–68°F) for the first 3 days. After this, you can let it rise in temperature up to 24°C (75°F) for the remainder of your 2 weeks, or until you have three identical gravity readings. Once you have decided what to let it rise to, don't let it fall. Otherwise, your yeast could flocculate and you'll have under-attenuated beer.

Bottle with 100g/3½oz of white table sugar to reach 2.1–2.3 volumes of CO_2. Age in the bottle for at least 2 weeks at room temperature. This beer will continue to develop with age.

Dubbel

Target Numbers:

Original gravity	1.066–1.068
Final gravity	1.004–1.008
ABV	7.8–8.2%
Bitterness	23 IBUs
Colour	70 EBC

Batch size (in fermenter)	20 l/qt
Expected efficiency	70%

Grain Bill

Pale Malt, Belgian	66.7% – 4kg/9lb
Wheat Malt	8.3% – 500g/1⅛lb
Cara–Munich Malt	8.3% – 500g/1⅛lb
Dark Candi Sugar	16.7% – 1kg/2¼lb

Hops

Hallertauer Mittelfrueh (4% AA)	First wort hop – 30g/1oz
Hallertauer Mittelfrueh (4% AA)	Boil 20 mins – 30g/1oz

Yeast

Belgian Abbey Yeast. For this, I'd go for Rochefort yeast (WLP540, Wyeast 1762), but you could also go for Wesmalle yeast (WLP530, Wyeast 3787) or Chimay yeast (WLP500, Wyeast 1214)
Alternatives: Dried Belgian yeast like Safbrew Abbaye or Mangrove Jacks Belgian Ale

Additional Ingredients

1 Irish Moss Tablet (such as Protofloc or Whirlfloc)

Method

Prepare your yeast. You'll need plenty. Clean and prepare your brewing equipment.

Bring 24 litres/quarts of water up to 69°C (156°F). Treat this water according to your water report.

Mash in. Maintain a mash temperature of 65°C (149°F) for 60 minutes.

Mash out – raise your grain temperature to 75°C (167°F).

Sparge with 4 litres/quarts of water at 75°C (167°F) to reach your pre-boil volume of no more than 23 litres/quarts.

Add your first wort hops and your sugar. Boil your wort for 75–90 minutes. Add your flavour hops at 20 minutes before the end of the boil.

Chill your wort to 18°C (64°F). Measure your original gravity. Liquor back with sanitary water to reach your intended OG.

Transfer your wort to a clean and sanitary fermenter. Aerate your wort and pitch your prepared yeast.

Ferment in primary fermenter at 18°C (64°F) for the first 2–3 days of active fermentation. Then, remove all cooling to let your temperature free-rise. Do not let it go above 26°C (79°F). Whatever temperature it reaches, keep it there until you've got three identical gravity readings. Expect this to take about 2 weeks from pitching.

Bottle with 120g/4¼oz of white table sugar to reach approximately 2.7–2.8 volumes of CO_2. This beer will benefit from quite a bit of bottle conditioning and will improve with age.

Saison (farmhouse ale)

Target Numbers:

Original gravity	1.058–1.062
Final gravity	1.008–1.010
ABV	6.3–6.5%
Bitterness	30 IBUs
Colour	7 EBC
Batch size (in fermenter)	20 l/qt
Efficiency	70%

Grain Bill

Pilsner Malt, Belgian	90.9% – 5kg/11lb
Wheat, unmalted	9.1% – 500g/1⅛lb

Hops

Saaz (4% AA)	First wort hop – 30g/1oz
Saaz (4% AA)	Boil 30 mins – 20g/¾oz
Saaz (4% AA)	Boil 15 mins – 30g/1oz

Yeasts

Saison Yeast, such as WLP565, Wyeast 3724 or Danstar Belle Saison
AND optionally one Brettanomyces strain. WLP670 American Farmhouse
blend is good and The Yeast Bay saison/Brett blend is truly excellent

Method

Prepare your saison yeast. Clean and prepare your brewing equipment.

Bring 24 litres/quarts of water up to 70°C (158°F). Treat this water according to your water report.

Mash in. Maintain a mash temperature of 64.5°C (148°F) for 90 minutes.

Mash out – raise your grain temperature to 75°C (167°F).

Sparge with 4 litres/quarts of water at 75°C (167°F) to reach your pre-boil volume of no more than 24 litres/quarts.

Add your first wort hops. Boil your wort for 90 minutes. Add your flavour hops at 30 and 15 minutes before the end of the boil.

Chill your wort to 18°C (64°F). Measure your original gravity. Liquor back with sanitary water to reach your intended OG.

Transfer your wort to a clean and sanitary fermenter. Aerate your wort and pitch your prepared saison yeast.

Ferment in primary fermenter at 18°C (64°F) for the first 2 days of fermentation. Then, stop any cooling to let your temperature free-rise. Once it's gone as high as it's going to get, heat to reach 30–32°C (86–90°F). Do not let the temperature drop until all activity has subsided – usually about 7–10 days. Once your yeast has flocculated, you can leave it for another week to condition if you want a clean beer. If you want an amazingly complex sour beer, add your Brettanomyces blend and leave for a month.

Bottle with 150g/5¼oz of white table sugar to reach approximately 3 volumes of CO_2.

East Coast IPA

Target Numbers:

Original gravity	1.066–1.070
Final gravity	1.008–1.014
ABV	7.2–7.8%
Bitterness	20 IBUs
Colour	9 EBC
Batch size (in fermenter)	20 l/qt
Estimated efficiency	65%

Grain Bill

Maris Otter	71.4% – 4.5kg/10lb
Oats, rolled	7.9% – 500g/1⅛lb
Wheat, unmalted	7.9% – 500g/1⅛lb
Munich Malt	7.9% – 500g/1⅛lb
Sugar, white	4.8% – 300g/10½oz

Hops

Centennial (10% AA)	First wort hop – 10g/⅓oz
Citra (12% AA)	Boil 20 mins – 10g/⅓oz
Amarillo (8.5% AA)	Aroma steep – 60g/2⅛oz
Simcoe (12.3% AA)	Aroma steep – 60g/2⅛oz
Citra (12% AA)	Aroma steep – 60g/2⅛oz
Centennial (10% AA)	Aroma steep – 60g/2⅛oz
Centennial (10% AA)	Dry hop – 30g/1oz
Amarillo (8.5% AA)	Dry hop – 40g/1½oz
Simcoe (12.3% AA)	Dry hop – 40g/1½oz
Citra (12% AA)	Dry hop – 30g/1oz

Yeast

Vermont Ale Yeast. Alternatively, use any characterful English Ale Yeast, and finish with US-05, WLP001 or Wyeast 1056 for dryness

Method

Prepare your chosen yeast – in this style, overpitching is a good idea. Clean and prepare your brewing equipment.

Bring 27 litres/quarts of water up to 69.5°C (157°F).

Mash in. Maintain a mash temperature of 64.5°C (148°F) for 60–75 minutes.

Mash out – raise your grain temperature to 75°C (167°F).

Sparge with 6 litres/quarts of water at 75°C (167°F) to reach your pre-boil volume of no more than 26 litres/quarts.

Add your first wort hops and sugar and boil your wort for 60 minutes. Add your hop addition at 20 minutes before the end of the boil.

Cool your beer to 75–79°C (167–174°F) and add your sizeable aroma hops. Steep these for 30 minutes at no higher than 79°C (174°F).

Chill your wort to 18°C (64°F), liquoring back with sanitary water to reach your intended original gravity.

Transfer your wort to a clean and sanitary fermenter. Aerate your wort and pitch your prepared yeast.

Ferment in primary fermenter at 18–20°C (64–68°F). After 3 days, ramp the temperature up as high as 25°C (77°F) for attenuation. Keep it there for a week. Make sure you have 3 identical gravity readings over 3 days before transferring. Transfer to your secondary fermenter and dry hop for 3 days.

Bottle with 110g/3⅞oz of white table sugar to reach 2.4–2.5 volumes of CO_2.

Session IPA

Target Numbers:

Original gravity	1.042–1.044
Final gravity	1.010–1.012
ABV	4.1–4.4%
Bitterness	40 IBUs
Colour	14 EBC
Batch size (in fermenter)	20 l/qt
Estimated efficiency	70%

Grain Bill

Pilsner Malt, German	78% – 3.2kg/7lb
Oats, rolled	5% – 200g/7oz
Crystal Malt	5% – 200g/7oz
Munich Malt	5% – 200g/7oz
Rye Malt	7% – 300g/10½oz

Hops

Centennial (10% AA)	First wort hop – 20g/¾oz
Centennial (10% AA)	Boil 10 mins – 20g/¾oz
Amarillo (8.5% AA)	Boil 5 mins – 20g/¾oz
Amarillo (8.5% AA)	Aroma steep – 60g/2⅛oz
Centennial (10% AA)	Aroma steep – 100g/3½oz
Mosaic (7% AA)	Aroma steep – 100g/3½oz
Mosaic (7% AA)	Dry hop – 100g/3½oz

Yeast

English Ale Yeast, such as White labs WLP002, Wyeast 1968 or Safale S-04

Additional Ingredients

1 Irish Moss Tablet (such as Protofloc or Whirlfloc)

Method

Prepare your chosen yeast. Clean and prepare your brewing equipment.

Bring 25 litres/quarts of water up to 71°C (160°F).

Mash in. Maintain a mash temperature of 66.5°C (151°F) for 60 minutes.

Mash out – raise your grain temperature to 75°C (167°F).

Sparge with 4 litres/quarts of water at 75°C (167°F) to reach your pre-boil volume of no more than 25 litres/quarts.

Add your first wort hops then boil your wort for 75 minutes. Add your fining tablet and chiller at 15 minutes. Add your hop additions at 10 and 5 minutes before the end of the boil.

Cool your beer to 75–79°C (167–174°F) and add your aroma hops. Steep these for 30 minutes at no higher than 79°C (174°F).

Chill your wort to 18°C (64°F), liquoring back with sanitary water to reach your intended original gravity.

Transfer your wort to a clean and sanitary fermenter. Aerate your wort and pitch your prepared yeast.

Ferment in primary fermenter at 18–20°C (64–68°F) for 2 weeks. Make sure you have three identical gravity readings over 3 days.

Transfer to a sanitary secondary fermenter and dry hop for 3 days.

Bottle with 120g/4¼oz of white table sugar to reach 2.5–2.7 volumes of CO_2.

Sour ale

Target Numbers:

Original gravity	1.034–1.036
Final gravity	1.002–1.004
ABV	4.2–4.4%
Bitterness	22 IBUs
Colour	4 EBC

Batch size (in fermenter)	20 l/qt
Estimated efficiency	70%

Grain Bill

Pale Malt, Maris Otter	68.6% – 2.4kg/5¼lb
Oats, rolled	31.4% – 800g/1¾lb

Hops

East Kent Goldings (5.5% AA)	First wort hop – 30g/1oz
East Kent Goldings (5.5% AA)	Boil 15 mins – 25g/⅞oz

Yeast

One Brett-saison blend, plus dregs from your favourite Bretty and sour beers. Good blends are available from The Yeast Bay, Omega Yeast, WHC or Bootleg Biology

Method

Prepare your chosen yeasts and dregs. You don't need to worry about pitching rates here. Clean and prepare your brewing equipment.

Bring 23 litres/quarts of water up to 69°C (156°F).

Mash in. Maintain a mash temperature of 64.5°C (148°F) for 90 minutes.

Mash out – raise your grain temperature to 75°C (167°F).

Sparge with 4 litres/quarts of water at 75°C (167°F) to reach your pre-boil volume of no more than 23 litres/quarts.

Add your first wort hops and boil your wort for 90 minutes. Add your hop addition at 15 minutes before the end of your boil.

Chill your wort to 18°C (64°F). Measure your original gravity and liquor back with sanitary water to reach your intended OG.

Transfer your wort to a clean and sanitary fermenter. This is still important, as we do not particularly want this beer infected with Acetobacter. Aerate your wort and pitch your yeast and dregs.

Ferment in primary fermenter at 18–20°C (64–68°F) for at least 2–3 months, or until your beer is smelling nice and funky and tastes very sour. It should have a good pellicle (Brett-crust) on top.

Bottle with 140g/5oz of white table sugar to reach roughly 3 volumes of CO_2. Don't use flimsy bottles.

California common ale

Target Numbers:

Original gravity	1.052–1.056
Final gravity	1.016–1.018
ABV	4.8–5.2%
Bitterness	37 IBUs
Colour	17 EBC

Batch size (in fermenter)	20 l/qt
Estimated efficiency	70%

Grain Bill

Pale malt, Maris Otter	90.9% – 4.5kg/10lb
Crystal Malt	5.1% – 250g/8¾oz
Amber Malt	4% – 200g/7oz

Hops

Hallertauer Mittelfrueh (4% AA)	First wort hop – 50g/1¾oz
Hallertauer Mittelfrueh (4% AA)	Boil 15 mins – 50g/1¾oz
Hallertauer Mittelfrueh (4% AA)	Aroma steep – 50g/1¾oz

Yeast

San Francisco Lager Yeast
(California Lager Yeast – WLP810, Wyeast 2112)

Additional Ingredients

1 Irish Moss Tablet (such as Protofloc or Whirlfloc)

Method

Prepare your chosen yeast. Make sure your yeast calculator is set to 'lager' – you will need lots and, likely as not, will need to make a starter. Clean and prepare your brewing equipment.

Bring 24 litres/quarts of water up to 71°C (160°F).

Mash in. Maintain a mash temperature of 65°C (149°F) for 60–75 minutes.

Mash out – raise your grain temperature to 75°C (167°F).

Sparge with 6 litres/quarts of water at 75°C (167°F) to reach your pre-boil volume of no more than 24 litres/quarts.

Add your first wort hops and boil your wort for 60 minutes. Add your flavour hops and finings at 15 minutes before the end of the boil.

Cool your beer to 75–79°C (167–174°F) and add your aroma hops. Steep these for 30 minutes at no higher than 79°C (174°F).

Chill your wort to 18°C (64°F), liquoring back with sanitary water to reach your intended original gravity.

Transfer your wort to a clean and sanitary fermenter. Aerate your wort and pitch your prepared yeast.

Ferment in primary fermenter at 14–18°C (57–64°F) for 2 weeks. Make sure you have three identical gravity readings over 3 days.

Bottle with 110g/3⅞oz of white table sugar to reach 2.4–2.5 volumes of CO_2.

Tripel

Target Numbers:
Original gravity	1.074–1.078
Final gravity	1.004–1.006
ABV	9.4–9.8%
Bitterness	38 IBUs
Colour	7 EBC

Batch size (in fermenter)	20 l/qt
Expected efficiency	70%

Grain Bill
Pale Malt, Belgian	72.6% – 4.5kg/10lb
Wheat Malt	8.1% – 500g/1⅛lb
Sugar, white	19.4% – 1.2kg/2½lb

Hops
Styrian Goldings (5.4% AA)	First wort hop – 40g/1½oz
Hallertauer Mittelfrueh (4% AA)	Boil 30 mins – 20g/¾oz
Hallertauer Mittelfrueh (4% AA)	Boil 15 mins – 30g/1oz

Yeast
Belgian Abbey Yeast, such as Wesmalle yeast (WLP530, Wyeast 3787), Chimay yeast (WLP500, Wyeast 1214) or Rochefort yeast (WLP540, Wyeast 1762)
Alternatives: Dried Belgian Yeast like Safbrew Abbaye or Mangrove Jacks Belgian Ale

Additional Ingredients
1 Irish Moss Tablet (such as Protofloc or Whirlfloc)

Method

Prepare your yeast. You'll need plenty. Clean and prepare your brewing equipment.

Bring 24 litres/quarts of water up to 69°C (156°F). Treat this water according to your water report.

Mash in. Maintain a mash temperature of 64.5°C (148°F) for 75 minutes.

Mash out – raise your grain temperature to 75°C (167°F).

Sparge with 4 litres/quarts of water at 75°C (167°F) to reach your pre-boil volume of no more than 24 litres/quarts.

Add your first wort hops and your sugar. Boil your wort for 90 minutes. Add your flavour hops at 30 and 15 minutes before the end of the boil.

Chill your wort to 18°C (64°F). Measure your original gravity. Liquor back with sanitary water to reach your intended OG.

Transfer your wort to a clean and sanitary fermenter. Aerate your wort and pitch your prepared yeast.

Ferment in primary fermenter at 18°C (64°F) for the first 2–3 days of active fermentation. Then, remove all cooling to let your temperature free-rise. *Do not let it go* above 26°C (79°F). Whatever temperature it reaches, keep it there until you've got three identical gravity readings over 3 days. Expect this to take about 2 weeks from pitching.

Bottle with 120g/4¼oz of white table sugar to reach approximately 2.7–2.8 volumes of CO_2. This beer will benefit from quite a bit of bottle conditioning and some age.

American pale ale

Target Numbers:

Original gravity	1.059–1.061
Final gravity	1.010–1.014
ABV	6.2–6.4%
Bitterness	48 IBUs
Colour	15 EBC

Batch size (in fermenter)	20 l/qt
Estimated efficiency	65%

Grain Bill

Pale Malt, US 2-Row	82% – 5kg/11lb
Munich Malt	8.2% – 500g/1⅛lb
Cara-Pils (Carafoam; Dextrine)	3.3% – 200g/7oz
Crystal Malt	3.3% – 200g/7oz
Melanoidin Malt	3.3% – 200g/7oz

Hops

Citra (12% AA)	First wort hop – 20g/¾oz
Citra (12% AA)	Boil 10 mins – 20g/¾oz
Citra (12% AA)	Boil 5 mins – 30g/1oz
Citra (12% AA)	Aroma steep – 80g/3oz
Citra (12% AA)	Dry hop – 50g/1¾oz

Yeast

Any English Ale Yeast, like Safale S-04, WLP002 or Wyeast 1968
Alternatives: Dry English Ale Yeast for a more sessionable beer, or go for
White labs WLP007, Wyeast 1098 or Mangrove Jacks m07

Additional Ingredients

1 Protofloc (Irish Moss) tablet

Method

Prepare your yeast. Clean and prepare your brewing equipment.

Bring 26 litres/quarts of water up to 69.5°C (157°F).

Mash in. Maintain a mash temperature of 64.5–65°C (148–149°F) for 60–90 minutes.

Mash out – raise your grain temperature to 75°C (167°F).

Sparge with 4 litres/quarts of water at 75°C (167°F) to reach your pre-boil volume of no more than 25 litres/quarts.

Add your first wort hops. Boil your wort for 60 minutes. Add your finings 15 minutes before the end of the boil. Add your hop additions at 10 and 5 minutes before the end of the boil.

Cool your beer to 75–79°C (167–174°F) and add your aroma hops. Steep these for 30 minutes at no higher than 79°C (174°F).

Chill your wort to 18°C (64°F). Measure your original gravity. Liquor back with sanitary water to reach your intended OG.

Transfer your wort to a clean and sanitary fermenter. Aerate your wort and pitch your prepared yeast.

Ferment in primary fermenter at 18–20°C (64–68°F) for 2 weeks. Make sure you have three identical gravity readings over 3 days.

Transfer to secondary fermenter and dry hop for 3 days.

Bottle with 110g/3⅞oz of white table sugar to reach 2.4–2.5 volumes of CO_2.

Index & suppliers

Index

Suppliers

These are simply the companies I have had excellent dealings with. None of them have given us any free stuff; they are here by their own merit.

THE MALT MILLER
– themaltmiller.co.uk
Rob sells me most of my yeast and malt, as well as a lot of my equipment and sundries. All of the buckets, taps and kegging equipment you see in this book came from him.

BREWUK
– brewuk.co.uk
Usually my first alternative when The Malt Miller is out of stock – a great selection of yeast and equipment, if a little pricey.

THE HOMEBREW COMPANY
– thehomebrewcompany.ie
These friendly Irish lads supplied all my kegs and, oddly, my first ever bottling stick. Thanks for helping me store my beer.

GLEN BREW, GLASGOW
– innhousebrewery.co.uk
My local. When I'm short of grain or in desperate need of a dried yeast I thought I had, this wee shop has saved my skin more than a few times.

HOME BREW BUILDER
– brewbuilder.co.uk
The UK's best source of all things stainless steel, shiny or otherwise unnecessary and expensive. I love it.

GET 'ER BREWED
– geterbrewed.com
Excellent hop and malt selection for the UK and Ireland, and great service too.

HOPS DIRECT, LLC
– hopsdirect.com
For when hop season starts, what better place to get awesome hops than direct from the growers? Get them fresh, use them quick.

SCREWFIX
– screwfix.com
Just down the road and plumbing fittings galore – they stock everything you need to build an entire brewery, if required.

Acknowledgements

I wouldn't be a beer fan, home brewer or commercial brewer without **Owen Sheerins**, who taught me everything in the beginning. Thanks too, to excellent brewers **Geoff Trail** and **Gareth Young**, who went through the book in search of brewing errors and found many.

This book wouldn't exist without the wonderful and so very, very patient **Stacey Cleworth**. Thanks to **Claire Rochford** for looking after the design, **Lucia Vinti** for her cover illustration and **Jonathan Baker** for typesetting. Thanks especially to **Sarah Lavelle** (you wonder). And obviously **James**, **Tilly** and **Elliott**, thank you so.

Every time, **Andy Sewell**, you amaze me. **Tim**, **Al**, **Lib**, looking back at these photos makes me wonder how you tolerated us, yet again. And **Will Webb**, our personal home-brewing and sourdough-baking designer, you have been amazingly tolerant.

My wife **Fenella** has supported me throughout these trying times, and I've a new addition to thank: **Lily Morton**, born 19th June 2020. Thanks as always to my family, who are there whenever they can be: **Mum**, **Dad**, **Magnus**, **Martha**, **Sandy**, **Dave**.

Thanks to **Drygate** brewery for letting us shoot there, and to **themaltmiller. co.uk** for the continued excellent service over the years.

Publishing Director **Sarah Lavelle**
Editor **Stacey Cleworth**
Head of Design **Claire Rochford**
Designer & Art Director **Will Webb**
Cover illustrator **Lucia Vinti**
Typesetter **Jonathan Baker (Seagull Design)**
Photographer **Andy Sewell**
Head of Production **Stephen Lang**
Production Controller **Nikolaus Ginelli**

First published in 2021 by
Quadrille, an imprint of Hardie Grant Publishing
Pentagon House
52–54 Southwark Street
London SE1 1UN
quadrille.com

Cataloguing in Publication Data: a catalogue record for this book is available from
the British Library.

9781787136977

Printed in China

DISCLAIMER: The recipes in this book were developed in metric quantities.
Imperial conversions have been added subsequently. The author **strongly** suggests
that you follow the metric measurements to achieve foolproof brewing results.